COMCASTED

How Ralph and Brian Roberts Took Over America's TV, One Deal at a Time

Joseph N. DiStefano

Camino Books, Inc.
Philadelphia

Manufactured in the United States of America

1 2 3 4 5 08 07 06 05

Library of Congress Cataloging-in-Publication Data

DiStefano, Joseph N., 1963–
  Comcasted: how Ralph and Brian Roberts took over America's TV, one deal at a time/Joseph N. DiStefano.—1st ed.
    p. cm.
  Includes bibliographical references.
  ISBN 0-940159-82-1 (alk. paper)
  1. Comcast Corporation—History. 2. Roberts, Ralph. 3. Roberts, Brian (Brian Leon). 4. Cable television—United States—History. I. Title

HE8700.72.U6D5 2005
384.55'5'06573—dc22   2004022138

Cover and interior design: Jerilyn Bockorick

This book is available at a special discount on bulk purchases for promotional, business, and educational use. For information write to:

Publisher
Camino Books, Inc.
P.O. Box 59026
Philadelphia, PA 19102

www.caminobooks.com

To my grandfathers

José N. Jané,
who solemnly enjoyed
the Spanish cable news,

and

Renato T. DiStefano Sr.,
who ran his TV with the sound off
and called it "the Idiot Box"

CONTENTS

Preface   ix
A Comcast Timeline   xi

Introduction   1
Can't Live without It ▪ "Like Don Corleone, without the Killing" ▪
A Gentleman, by Comparison ▪ Comcast and Its Critics

1   Foundations   11
In the New World ▪ Philadelphia ▪ Her Fortune, His Future ▪
Birthright ▪ The Family Way ▪ Bent Putters

2   Salesman   21
The Ad Man ▪ The Big-Government Businessman ▪
Ralph in Charge ▪ The Road Not Taken

3   Pioneers   27
Live from Philadelphia ▪ "The Greatest Thing Since Stealing" ▪
Made in Pennsylvania ▪ Wired ▪
The Bare-Chested Antenna Salesman ▪ The Big Fish

4   Deep South   37
"Boom, There's Explosions" ▪ Hiding from the Customers ▪
Useful People ▪ Cable Goes to Town ▪ Slicing the Pie ▪
Ralph in Doubt? ▪ Band of Brothers

5   Capital   47
To Sell and to Hold ▪ Giving It Away ▪ The Way We Live Now ▪
The Adopt-a-Nation Vacation ▪ Gentleman, Monopolist, Bear

6   *Satellite*   53
Dirty Movies ▪ Look, Up in the Sky ▪
County Fairs and Heavyweights ▪ Gold Rush ▪ Rent-a-Citizen

7   *Spreading Out*   59
Half for Me? ▪ The Also-Rans ▪ Freeing the Fortune 500 ▪
Other People's Money

8   *Muscles*   65
Crown Prince ▪ Growing into It ▪ Show the Boss How ▪
"We've Been Comcasted" ▪ Close to Home

9   *Controlling Interests*   73
"Ten Times as Rich" ▪ All for Brian ▪ Family Men ▪
The Space Shuttle and the Titanic

10   *Living Color*   79
Bang-Bang vs. Boom-Boom ▪ Urban Warriors ▪
Two Sets of Books ▪ What You Can't Get

11   *Comcast Country*   85
"Sinister and Diabolical" ▪ Let the Feds Do It ▪
The Biggest Piece ▪ "What's Next?"

12   *Shopping*   91
Song and Dance ▪ Soft Sell ▪ "It's Shopping, and It's TV" ▪
Nothing Personal ▪ Air and Wire ▪ Hanging Up

13   *Sports*   101
Owners, Not Fans ▪ Bullies on Broad Street ▪
So You Want to Own a Pro Sports Team ▪ The Real Owner ▪
Come and Play

14   *Microsoft Buys In*   109
"As Monopolistic as You Can" ▪ Learn and Do ▪
What Competition? ▪ Message from Bill ▪ Under the Gun ▪
10 Percent

15 *Market Power* 117
Your Money, Our Choice ■ "Widespread Dissatisfaction" ■
Get Outta Town ■ Doesn't Like to Share

16 *Doing Unto Others* 125
Charity Muscle ■ Not Giving Much Away ■ Comrades ■
Empty Nesters, Passive Heirs

17 *Friendly Powers* 133
Trade Show for George Bush ■ America's Mayor ■ Not the News ■
Senator Lott's Constituents ■ Two Faces on K Street ■
Wiring the Capital ■ Lights Out

18 *Beating Ma Bell* 147
King and Kingmaker ■ Strong-armed ■ More for Less ■
"Don't You Dare Give Up" ■ The Bigger They Come ■
Smokescreen ■ What They Paid For

19 *Comcast's Critics* 159
No Sense ■ Monopoly in Motion ■
Who's Paying for This Microphone? ■ On the Cheap ■
Consumer Outrage ■ "Bad for Business?"

20 *Beyond Disney* 167
Who Wants a Price War? ■ What Murdoch Didn't Have ■
Beyond Reality ■ "A Logical Next Step" ■ You're Buying What? ■
"Highly Underwhelming" ■ Wishful Thinking ■ Hello, Operator ■
Too Soon the Prophet ■ Vote Early and Often ■
Spammer, Brawler, Thieves ■ The Tony Soprano of Cable? ■
How Cable Should Have Been ■ Shame on Us

Notes   187
Selected Bibliography   206

PREFACE

While more famous moguls rose and fell, Ralph Roberts, joined by his son Brian, spent forty years building Comcast Corporation into the nation's top merchant of TV news, sports, movies, and pornography. They cultivated powerful people in government, finance, and technology; launched sneak attacks on would-be competitors; assembled a chain of local monopolies linking most of the nation's major metro areas; and used their clout to win lucrative rates from the entertainment industry, advertisers, and the viewing public. Lately, Brian has bet the company's future on the Internet and video-on-demand, and launched takeover attempts against some of America's biggest and best-known companies. *COMCASTed* tells their story.

Ralph and Brian Roberts declined to be interviewed for this book. Among reasons cited by their aides: Ralph is considering his own memoirs; Brian values his privacy; and neither the author nor his publisher would agree to prior review. Comcast did make Ralph's longtime chief financial officer, Julian Brodsky, available for a couple of chaperoned interviews on historical subjects, and spokesman Tim Fitzpatrick said the company would not object to the use of several photos it had made available to news media; but no one at Comcast presumed to speak for Ralph or Brian.

However, the student of Comcast has recourse to a vast literature. When it has suited their purposes, Ralph and Brian have given interviews to many reporters, including the author on a couple of brief occasions. Ralph Roberts, a veteran salesman and sometime advertising man, knew early the importance of telling a good story about himself, and has told many to audiences, interviewers, peers, and rivals, though some have changed with the years. As Brian's profile has risen with his business audacity, he has resorted to investors' conference calls to get his message out; the calls are by invitation only, but the public can listen in, and transcripts are readily available. And, since the cable industry has been regulated or at least monitored by federal and municipal governments from its early days, and Comcast has been a publicly traded company for a generation, libraries of data and narrative on Comcast's ambitions, self-justifications, and self-defenses have accumulated over four decades—as have responses from boosters, critics, and rivals.

This book cites many such accounts in the endnotes; it also refers to books on the early years of cable, including the published memoirs of the late Comcast cofounder Daniel Aaron, a onetime reporter for the *Philadelphia Bulletin*. Comcast's David L. Cohen recommended and lent the author copies of Stephen A. Keating's *Cutthroat* and Kathi Ann Brown's *Wired to Win*, which, with other contemporary works, supply useful background on how the freewheeling, family-owned cable TV business grew and coalesced into the current handful of media giants.

The author is indebted to a number of Comcast-watchers for their generosity and patience with his questions. They are credited in the text and notes, when they have allowed their names to be used: for example, among Ralph Roberts's onetime rivals, Harold Fitzgerald "Gerry" Lenfest, whom Ralph disposed of as a competitor by turning him into a billionaire philanthropist; and, of former Comcast employees, the relentlessly positive Pat Croce, who learned the wide freedom and abrupt limits of serving under Roberts family rule. Many who encountered Ralph and Brian during key periods, like Ed Breen, now chief executive of Tyco International Limited, and Ted Aronson, one of the most successful of the current generation of Philadelphia-based professional investors, supplied insights and anecdotes, as did Comcast critics in media and labor advocacy groups.

At the author's direction, Cousin Bob Foley of WriteFocus Communications in Cortlandt Manor, New York, dug up documents and other records that shed light on the Roberts family's early life in this country and before. Among professional colleagues who lent their support and guidance, the author is especially grateful to Peter Binzen, author, editor, columnist, and elder of the Philadelphia business press, for his suggested improvements to the original manuscript, and for introducing him, at Philadelphia's Franklin Inn Club, to Camino Books publisher Edward Jutkowitz, who commissioned this book. Michael Panzer and his overworked news research staff opened their archives and made possible the purchase of photos. And editor Barbara Gibbons showed patient grace in the long process of improving the text.

The author especially thanks his sweet wife, Linda, who provided the most encouragement of all; Nick, John, Mark, Carl, James, and Maria, who kept asking when Dad's book was gonna get finished; and his parents, Ren and Marta, for all the Saturday mornings when they turned off the TV at 9 A.M. and brought the kids outside to discover their world.

A COMCAST TIMELINE

1904–1933 Ralph Roberts's grandparents move from Russia to New York with his dad; Ralph's father takes American name Robert Max Roberts. Sells patent medicines; develops pharmacies; invests in real estate; supports Jewish immigrant charities. Ralph Joel Roberts, born **1920**, middle child of three. Family moves to suburban New Rochelle, **1925**. Bob Roberts dies in **1933**, leaving family in reduced circumstances.

1937–1942 Ralph moves to Philadelphia with his mother, who dies soon after. Works his way through University of Pennsylvania's Wharton School; graduates, **1941**; commissioned, U.S. Navy; marries Suzanne Fleisher, daughter of prominent Philadelphia German–Jewish family, in **1942**; spends war at Philadelphia Naval Base.

1945–1946 Ralph develops and promotes "Bob Hope Centric Putter," without Hope's permission, at East Coast golf courses. Abandons product after it proves defective.

1946–1948 Ralph works for Aitkin Kynett ad agency in Philadelphia. Recruited by pioneering consumer marketer and liberal U.S. Sen. William R. Benton, D-Conn., to help sell canned music (Muzak) and write political ads opposing conservative McCarthyism.

1948–1950 Ralph works at Benton's Muzak Corporation, selling prerecorded music for elevators and waiting rooms.

1951 Ralph joins Pioneer Belt Company as ad director and potential successor to ailing owner.

1951–1961 Backed by Suzanne's family and connections, Ralph takes control of Pioneer. Expands product lines; is accused of copying rivals' designs. Suzanne active in Philadelphia politics and public television. Five children born to Ralph and Suzanne; fourth is Brian Leon Roberts (**1960**), the only one to take much interest in Ralph's business.

1953 As cable TV expands, led by Philadelphia antenna-maker and cable financier Milton Jerrold Shapp, Philadelphian Warren "Pete" Musser's investment firm acquires a cable system in Tupelo, Mississippi. Musser would later involve Comcast in television programming (via the highly profitable QVC) and Internet investments (through dot.com meteor, Internet Capital Group).

1961 Ralph sells Pioneer Industries and goes looking for less competitive businesses. Assembles a grab bag of companies under the title International Equity Corporation.

1963 Investor Musser seeks buyer for 1,200-subscriber Tupelo cable system. Broker Daniel Aaron matches seller with Ralph, who hires Aaron and accountant Julian Brodsky for his new American Cable Systems division, with backing from the Fleishers and their Philadelphia connections.

1963-1969 American Cable builds and buys cable systems across Mississippi and the U.S., competing with hundreds of corporate-backed and family-owned systems for local cable monopolies.

1969 International Equity renamed Comcast Corporation; includes in-store broadcasts and recorded background music, as well as cable TV.

1972 First public share offering, stock traded on the new Nasdaq stock market. Price soon falls below $1.

1975 TV satellites make possible HBO, the first premium cable network.

Late 1970s Sports and adult movies give cable an edge over broadcast TV in urban markets for the first time. Big corporations and private investors pour billions into cable companies.

1981–1988 After graduating from Wharton, Brian is promoted from controller in Trenton, to assistant general manager in Flint, Michigan, then to general manager back in Trenton. He promotes pay-per-view from South Africa despite the anti-apartheid ban. In **1984**, Brian is named vice president for operations of Comcast Cable; four years later, as executive vice president—married, with the first of his three children on the way—Brian joins the Comcast board. Comcast established as an investor in British cable.

1984 After a twenty-year struggle, Comcast wins the most lucrative cable TV franchise in Philadelphia and moves its headquarters downtown. Expands in suburban Baltimore, Detroit, and other key markets. Worried about competition, Ralph Roberts tells cable operators they must find, control, and charge for services that antenna and satellite-dish users won't be able to get.

1986 Comcast joins three other cable companies in splitting up Westinghouse cable division; the move doubles Comcast's customer base to over one million. Comcast invests in start-up, home-shopping network QVC.

1988 Comcast buys half of Storer Communications and becomes the No. 5 U.S. cable company, with two million customers. Hires lawyer Lawrence S. Smith as chief of finance and administration; his penny-pinching balances Brian's ambitions. Brian envisions a wireless empire; Comcast buys cell-phone systems in New Jersey and Delaware.

1990 Brian Roberts named president of Comcast; Ralph still in charge. Banker John R. Alchin joins top Comcast management.

1992 Comcast buys Metrophone cell business, giving the company three times as many phone customers as cable customers.

1993 Brian recruits Hollywood mogul and Fox Network founder Barry Diller as head of QVC. Over the next two years, Diller tries to use QVC to take over Paramount Studios and CBS, threatening Comcast's control over the home-shopping network.

1994 Comcast buys Maclean Hunter's U.S. cable operations, making Comcast the No. 3 U.S. cable operator, with over 3 million customers.

1995 Backed by Brian's sometime mentor John Malone (TCI) and New York investment bankers, Comcast outwits Diller and takes outright control of QVC. Comcast also buys E.W. Scripps cable, with nearly 1 million customers. Brian becomes chairman of the National Cable Television Association.

1996 After the death of Malone's partner, Brian Roberts joins Microsoft in a secret attempt to take control of Malone's company, TCI. Malone is outraged; Roberts later apologizes. Comcast persuades NHL Flyers' owner to give up controlling interest in the team; takes control of NBA 76ers and gets the city's largest bank to pay millions of dollars for naming rights for a new arena; forms Comcast SportsNet, which controls indoor venues and Philadelphia pro sports TV, and keeps these programs off rival satellite TV.

1997 At a meeting of cable executives, Brian Roberts challenges Microsoft chairman William H. Gates Jr. to invest in cable companies if he wants their business. Gates invests $1 billion in Comcast. Comcast takes control of E! Entertainment channel in deal with Walt Disney Company.

1998 Ralph informs owners of Jones Intercable he is taking over the company and its million-plus customers. Pressed by competition from Rupert Murdoch's British Sky Broadcasting Group, Comcast sells its U.K. cable interests. Former Disney executive Stephen B. Burke joins Comcast as head of cable.

1999 Stung by strong competition and weak profits, Comcast sells its cell phone business. Announces a string of small cable mergers but is rebuffed by suburban Philadelphia's Lenfest. Comcast announces $60 billion merger with Media One but is outbid by AT&T. As consolation, Comcast gets $1.5 billion plus rights to 2 million AT&T subscribers—including newly acquired Lenfest.

2000–2002 Comcast trades cable systems in several states with AT&T and Adelphia to consolidate its hold on key markets. Tries to buy AT&T cable systems; AT&T refuses. Comcast hosts future President Bush, Sen. Trent Lott, and other Republican notables at GOP national convention in Philadelphia; in **2002**, hires convention co-chair David L. Cohen, a prominent Philadelphia lawyer, to head political and PR work.

2001 Comcast adds sports network and cable systems in Washington, D.C.–Baltimore area, plus two national minor-sports channels. Begins high-definition TV; prepares video-on-demand, game channel. AT&T agrees to sweetened Comcast offer; Comcast eventually pays $51 billion for systems AT&T spent twice as much to acquire. Comcast more than doubles, to 21 million customers.

2003 Comcast sells QVC for nearly $8 billion. Announces African American–themed and Chicago-sports networks; promises to improve Internet phone service. Comcast Internet sign-ups rise sharply; stock rises to No. 4 in value on Nasdaq exchange. Comcast gives Pennsylvania Gov. Ed Rendell a sports talk show; uses wife of Maryland Gov. Robert Ehrlich as lobbyist. FCC Chairman Michael Powell promotes Comcast services. Brian begins secretly exploring an offer for Walt Disney Company.

2004 Convinced by his advisers, or his own wishful thinking, that Walt Disney Company's board would consider an offer, Brian Roberts bids $66 billion. The board unanimously rejects Roberts's first attempt, and Comcast shares fall amid speculation on its goals, prospects, and likelihood of a better offer. Roberts declares intention to expand phone service, video-on-demand. Gains video playback rights to SONY, MGM, NFL programs.

2005 Comcast reports rising sales as affluent customers prove willing to pay far more than $100 each month for combinations of Comcast cable, Internet, video-on-demand, and phone services.

Sources: Securities and Exchange Commission, Form 425 filing, 11 February 2004, and other Comcast documents; author research.

INTRODUCTION

Can't Live without It

Americans mostly don't get to choose which cable TV peddlers bring news, sports, movies, games, and pornography into their homes, and they've been measurably annoyed by the result: rising fees, spotty service, too many ads, too many channels no one wants. And now all those offers—by phone, mailbox, e-mail—pushing fancy new digital media stuff.

But when Brian Leon Roberts, head of the nation's largest cable monopoly and its fastest-growing Internet service, wants to explain himself, he doesn't go before cameras and talk to his customers. No; like other modern CEOs, the second-generation chief of Comcast Corporation goes before a screened audience of his biggest fans, professional investors,[1] and tells them things he hopes they want to hear: how using tens of billions of dollars of their money to buy AT&T or Walt Disney Company will make them all richer; and how, despite the threat of new competition and the promise of improved technology, he sees no need to stop boosting prices for his tens of millions of customers, most of whom he acquired quite recently, without asking their permission.

To those investors and reporters, like the author, who, given a chance, pester him with the usual questions about cost and competition, Brian is ready with a level reply: "You're connecting the wrong dots." Or, "That's not our message this year."[2] His eyes are looking ahead, at everyone's future consumption of digital media—how he's going to make it

1

pay, and how he's going to use it to hold his own against global media buccaneer Rupert Murdoch, satellite, wireless, the phone companies, and the programmers and the government, if need be. You may not share his vision, but Roberts is betting that your kids are on his side.

Creating a giant corporation is a real achievement. Roberts and the men and women hired to do his bidding impress close observers with their hard work and their tight focus. But to what end? They have used their talents and energies in the service of monopoly. Unlike Murdoch, whose insurgent Fox Networks and Sky Broadcasting challenged fat, entrenched corporations and gave consumers real choice, the Robertses have sought to avoid competition, seeking only businesses where they can charge high fees to customers who have few or no alternatives.[3]

In early 2004, the nation's pro investors—pension funds and mutual funds and money managers to the rich—voted Roberts America's "Best Chief Executive." But in customer service polls, Comcast ranked poorly.

Who is this, who can so cheerfully and profitably repeal the laws of supply and demand, charm Washington, intimidate Hollywood, deny the Philadelphia Flyers and 76ers to satellite-dish TVs, and bend the viewing public to his will? Brian's takeovers, his broad and deep attention to detail, and his pricing philosophy—charge the people more, because you can—are all made possible by a chain of monopolies in Comcast's basic product, cable TV, which now extends to seventeen of the nation's twenty biggest markets. It was handed down from his father, Ralph Joel Roberts, who spent four decades preparing Comcast to be the nation's largest cable company, and preparing Brian to inherit it and increase its dominance of U.S. video.

Back when Brian was still in kindergarten, Ralph Roberts's first employee called cable "the greatest thing since stealing."[4] Today, with everyone who wants cable already wired and others choosing cheaper satellite connections, Brian has been maneuvering to extend his father's success by similarly dominating the Internet and the market for video-on-demand—the television programs you want, when you want them, for a fee. To gain this business he mounted audacious raids against members of the nation's most exclusive corporate club, the thirty blue-chip stocks that make up the Dow-Jones Industrial Average. In 2001 he raided AT&T; defying the skeptics, he boosted sales and cut Comcast's towering debt; and then, on the brink of making Comcast solvent, or at least cash-flow positive, in 2004 he went after Walt Disney Corporation, which, however, ignored him.

By the time Roberts tried to get Disney chief Michael Eisner's job, Comcast was worth more than almost any other company listed on the Nasdaq stock market, with the exception of software giant Microsoft

Corporation. Not by coincidence, Microsoft was Comcast's biggest investor, though Ralph Roberts had made sure to arrange things so his son Brian, and not Microsoft chief Bill Gates, had special voting powers and full control.

Yet media is a volatile business. Less than a year before Brian won top honors in the CEO beauty pageant, Comcast's value, and with it the Roberts family fortune, went skidding on news that the big regional phone companies had finally gotten their act together and were starting to offer cheap rival Internet service. Former phone companies like Verizon were hitting Comcast where it hurt. Comcast had fought for years to avoid competitive pricing for cable TV. But now, in his lucrative Internet business, Roberts was haunted by the promise of the most basic feature of a free-market economy: a price war. By knocking Comcast's share price off its high, investors were showing they believed Comcast was vulnerable to cut-rate competition. For customers, competition would mean choice and saving money. But to Roberts, the very idea of price competition was moronic; it seemed suicidal. It wasn't part of how his company was used to seeing the world.

"I can't understand the reaction, violent as it's been," Roberts told a crowd of investors who gathered at Manhattan's Plaza Hotel in early 2003. (In Brian's world, this threat was perceived as "violent.") "I don't understand why you need this lower price." Forget about what things cost, he urged the crowd. Keep your eye on the product. "We're talking about increasing speed, increasing reliability." Who wouldn't pay more for that? he asked.

Who, indeed? Comcast had lately forced itself on 12 million American families by taking over their cable company, AT&T. Big mergers are messy. Comcast faced labor unrest in Pittsburgh and Maryland; lawsuits by angry towns in New England, the Midwest, and the West Coast; and general consumer outrage at the company's seemingly automatic price increases, in merry contrast with the stable prices of food, clothing, and other necessities. But Brian and his team had prepared for all of that. Not only was Comcast well positioned to bargain down programming fees (though it let other cable companies lead that fight in public), even as it boosted advertising rates. Comcast was also poised to exploit the benefits of the cable industry's hard-won lobbying victories, which had mostly freed it from irritating local regulation, while permitting the company to block cheaper satellite TV systems and other would-be competitors from tuning in pro sports programs Comcast controlled in its home market of Philadelphia.

With the Internet, Comcast and its cable colleagues again enjoyed special advantages: While federal law forced rival Net providers like the phone companies to share their systems as if they were public highways, Comcast had no legal obligation to share its 20 million home cable hook-ups with anyone except its chosen customers. As a result, Brian was able to boast that, even as AOL and other big Internet providers were cutting prices and still losing business, Comcast could boost its stand-alone high-speed Internet charge from $40 to $45 a month—five times the rate of inflation—and still manage to sign up millions of famished Net crawlers through its local cable monopolies.

In Congress, U.S. Sen. Ron Wyden had just compared Comcast to a restaurant that "force-fed" five-course dinners at outrageously high prices to hungry families that just wanted burgers.[5] But Wyden was a liberal Democrat; so were other leading Comcast critics, like the American Civil Liberties Union and the Communications Workers of America. Ralph Roberts had worked for liberal Democrats, married one, and hired more than one when he was starting out in business; but such views hadn't really mattered for years, in business or in Washington or in the minds of most voting Americans. Having branded and hosted George W. Bush's nominating convention and hired the brightest media lobbyists money could buy, Comcast had friends and admirers in Congress, where Senate Majority Leader Trent Lott called both Robertses his "constituents"; at the Federal Communications Commission, where Chairman Michael Powell promoted new Comcast services while ignoring earnest critiques of the company's pricing power; and in the governor's mansions of states like Maryland, whose First Lady was a Comcast lobbyist, and Pennsylvania, where the governor himself hosted a Comcast talk show.

To explain Comcast's success at pushing its bills on Middle America, Brian Roberts used metaphors of childlike dependency, or worse. High-speed Internet is like a narcotic, or like TV itself, he told the New York crowd: "Once you have it, you can't live without it, it's so transforming of a product."

But in a free market, aren't prices supposed to drop over time? Doesn't success attract competition? Won't Comcast, like other monopolies before it, be forced to suffer price cuts and reduced profits? he was asked.

"Not today. Not tomorrow," Roberts insisted in his mild Philadelphia accent—not the nasal Main Line drawl, but the urban-white-ethnic variant, which sounds like a cold. "Plenty of industries have gone down the road of price wars," Brian warned. But Comcast has no desire to follow them.

"The trick," he told the crowd, "is not to let yourself become a commodity."

Once more for emphasis: "We're not a commodity."

"Like Don Corleone, without the Killing"

They dominate American TV, yet until their Disney bid made them front-page news the Robertses remained less recognizable than such loudmouths as CNN founder Ted Turner, cable visionary John Malone, or global media mogul Rupert Murdoch. Ralph and Brian are typically courteous, and generally bland, and seldom controversial in public. Ralph is witty and polished; Brian deeply competitive but disarmingly gawky. They marry for life, and raise mostly successful children, quietly.

Still, those who've worked with and against them and seen them under pressure have formed very strong impressions. "They're like the Corleones," said Pat Croce, former front man for Comcast's Philadelphia 76ers, comparing the Roberts blood and business families to the loyal, honorable, decisive, utterly fictional Mafia dons depicted in Mario Puzo's cynical book and Francis Ford Coppola's prettied-up movies. Of course, Croce meant this as a compliment: "I love Ralph Roberts; he's Don Corleone, without the killing."[6]

A Comcast rival prefers to compare Ralph and his wife of more than sixty years, the former Suzanne Fleisher of the Philadelphia Fleishers, with those other dynasts, Joe and Rose Kennedy: "She had the family background, and he was the devilish entrepreneur." Stretch the comparison, and you might weigh fortunate son Brian, with his wonderful cable tollgate connecting Hollywood, Washington, Wall Street, Silicon Valley, and your living room, alongside the late John F. Kennedy, who merely became president.

Both fatherless when they met, Ralph and Suzanne completed one another. Her family money and connections fed his business career, ending the wrenching moves, the name changes, the personal re-invention that had characterized his own immigrant family after it fled the Jewish edge of the old Russian empire. And his cheerfully ingenious aggression toward life freed her, converting the burdens she felt from her own exalted yet stifling background in Philadelphia's century-old German-Jewish elite, into a privilege for enjoying the world.

So what have heredity, environment, borrowed money, carefully cultivated connections, and relentless opportunism brought the Robertses? Not the desire to purchase immortality through good deeds. Though the family has spread its money widely in small doses, Philadelphia philanthropies have yet to drag a really groundbreaking, carve-it-in-stone donation from the most prominent of the region's active business dynasties. What, then, have they gained? Watch Ralph with his small smile and dotted bowtie and Suzanne in her bright clothes stroll out of their hotel suite near Philadelphia's Rittenhouse Square; share lunch at the diner down the block; separate so he can walk to his office to admonish and admire his son, successor, and heir, while she tapes five-minute segments of her TV show, slotted daily on Comcast's droning local programming channel; then resume each other's company at their custom Colonial-style estate in Brandywine horse country, within a knot of their locally raised grandchildren, hard by the Dorrances and du Ponts, who lead the Philadelphia branch of the older billionaire industrial aristocracy. Taking it all in, an observer may come to suspect it might actually be possible for a fortunate and canny couple to have it all in this life—thanks, in this case, to American television viewers' endless willingness to pay ever higher prices to watch the always flickering screen, instead of going out to make something grand of their own lives.

A Gentleman, by Comparison

Moving sometimes patiently and sometimes briskly among shouters and maniacs, they kept confident and cool. Yet Ralph Roberts and his lieutenants were never exactly secretive about their broad business goals. They have, over the years, developed a cycle of stories to tell reporters, business contacts, industry crowds at awards banquets, lawmakers they have cultivated, and other mostly sympathetic audiences. Sometimes details change, or evolve in the retelling. Many of the stories are comically self-deprecating; generally they reinforce the company self-image of steady focus, aggressive patience, careful planning, and fortunate timing. They don't make explicit what runs to the heart of Comcast history: the deep vein of opportunism; the willingness to look past a range of evils in pursuit of a deal or an advantage; and the keen ability to represent themselves as whatever the customer, regulator, investor, or politician who could help them needs to see. *Forbes* magazine isn't the only observer that has called Ralph and Brian wolves in sheep's clothing.[7]

Ralph Roberts didn't care particularly about television, or building a fortune just to give it away. He wanted to build an empire "and hand Brian a nice thing," recalled his top lieutenant, Julian Brodsky. Ralph's motivation was "pragmatic: We wanted to build a great company. We had to be opportunistic. We didn't have any goal other than the next deal."[8]

His first business venture as an adult was to sell celebrity golf clubs—without the permission of the national celebrity whose name he usurped. When the clubs, once sold, proved defective, Ralph walked away from the business and the people who had trusted him. In another favorite story, Roberts and his aides won a key early contract by promising to meet impossible deadlines, then shutting their local office and making their service staff scarce so they could truthfully report no applicant was waiting to be hooked up. But unlike the contemporary companies that built systems for the tax and sale advantages, Comcast, once it wired a neighborhood, was there to stay.

Much later, hard-charging John Malone, of Tele-Communications Incorporated, who had mentored Brian, accused him of repaying his trust by using the death of Malone's business partner to join secretly with Microsoft in a scheme to take over TCI.[9] But with the scheme exposed, Brian apologized—it cost him nothing—and they're friends again, swapping assets and bracing for Rupert Murdoch's threatened price assault via satellite TV. Ralph and Brian launched more successful raids against old peers, like the Lenfest and Jones families, cornering them into deals before bothering to let them know.

Comcast claims a progressive record on affirmative action and promoting women and minorities; Time Warner chief Dick Parsons, who is black, has felt comfortable enough to tell Jewish jokes on Brian in public. Yet both Ralph and Brian started their cable careers by exploiting situations others wouldn't touch because of explosive racial conflicts. Ralph cultivated the small-town power structure in white supremacist Mississippi in the 1960s. In time, he made friends with influential old segregationists like Sen. Trent Lott, who supported Comcast's regulatory agenda and was guest of honor at one of Comcast's bigger parties, the Republican convention in 2000. Similarly, in the 1980s, one of Brian's first initiatives as a Comcast manager was a pay TV promotion deal at a whites-only South African resort, ignoring a worldwide boycott of the outcast apartheid state. At that very time, Comcast was also offering sweetheart investment deals to African Americans in Philadelphia to help win a neighborhood monopoly from the black-led city council. It wasn't about black and white. Comcast cared about the green; it was equal-opportunist.

Comcast executives would boast to potential investors of huge operating profits, while telling the federal government the company was barely making money, so it could legally avoid federal taxes. Further down the chain of tax avoidance, Comcast opened nominal offices for hundreds of subsidiaries in a Delaware office hotel, enabling itself to avoid taxes to states where the company had customers. Ralph, and later Brian, would travel south by Amtrak Metroliner and promise Congress or the FCC that what was good for Comcast and its cable allies was good for competition and consumers, then train north to New York and tell brokerage analysts how Comcast planned to keep consumer rates high by avoiding competition and squashing, not just individual competitors, but whole classes of competitors.

In late 2002, Ralph Roberts stood before celebrity interviewer Charlie Rose to collect the Media Institute's yearly award for "leadership in promoting the vitality and independence of American media and communications." Soon after, Roberts's company faced a formal protest for its "apparent desire to sanitize an expression of political speech" when it canceled antiwar ads opposing the policy of the administration in Washington.

Ralph was initially squeamish about that cable staple, R-rated movies—the very product that made cable profitable. He was also "against the Playboy Channel," noted Julian Brodsky. But when customers clamored for sex shows, "he came around," Brodsky added. "Times change."[10]

Brian showed his own detachment from the matter in which he trades when he purchased pro sports franchises, not as a direct expression of his ego, but as a business move toward local programming monopoly. He rarely haunts the locker room or visits the stadium to watch his employees, who include star guard Allen Iverson (Brian earns more). Personally, he prefers no-prisoners squash, golf, and skiing, individual sports where he can smash his opponents, including lesser Comcast executives as well as big-name business allies and rivals.

Through it all, Ralph Roberts has managed to enjoy, and Brian to acquire, the reputation of a gentleman, the genuine nice guy who, though a tough and utterly serious negotiator, never raises his voice, avoids injecting business into social situations, doesn't lower himself to insults, trusts his subordinates, and keeps his ego out of the way of his business and himself mostly out of the public eye. In his memoirs, Reed Hundt, the Washington lawyer who was Bill Clinton's top telecom regulator, called Ralph a "modest" and refined fellow whose harshest gesture was to raise his silvered eyebrows;[11] Roberts's competitors at AT&T and Time Warner have paid similar compliments. "A kindly gentleman," Hollywood mogul Barry Diller told *Fortune*—even after recalling how Ralph and Brian had broken their word and forced him out of a job.[12]

Gentlemen—with that history, in this business?

By comparison, yes. Like other big companies, Comcast has been known to break rules from time to time. But its officials weren't convicted of bribery and imprisoned during cable's freewheeling growth years, as too many of their more and less prominent competitors were. And Brodsky could boast the company was never forced to restate earnings because of accounting irregularities—again unlike many competitors. In the end, if Comcast has emerged from the "franchise wars" and

the sudden-death merger battles as the most successful of the serial monopolists and cynical promoters who mass-market so many mindless
and exploitative programs, it only goes to show that nicer guys sometimes have more staying power.

Ralph Roberts succeeded without the bluster of News Corporation
chairman Rupert Murdoch or CNN founder Ted Turner, without AT&T
boss Michael Armstrong's multibillion-dollar blunders, or brilliant TCI
founder John Malone's self-defeating provocations (Malone once called
for the death of then FCC Chairman Hundt, though he later took it
back); also without cable pioneer Irving Kahn's public-corruption convictions, or the corporate-fraud criminal indictments that ended the career and reputation of Adelphia Corporation founder John Rigas. The
family has likewise avoided the annoying public arrogance associated
with famously successful monopolists like Microsoft's William H. Gates
or, for that matter, Standard Oil's John D. Rockefeller.

So by the relative standards both of the cable industry and of monopolists, that makes the Robertses, father and son, the very image of
gentlemen. Statesmen, even.

That reputation has been one of their most useful assets.

Comcast and Its Critics

Not everyone appreciates the way the Robertses have fashioned American media for their own purposes. Consumer activist Jeff Chester, the
Ralph Nader for digital America, called Comcast "a threat to competition, choice and democratic discourse"—and to Internet access.[13] Comcast charges some of the cable industry's highest prices, but respected
researchers from the University of Michigan and J. D. Power & Associates last year ranked Comcast and its new AT&T properties at or near
the bottom of the customer-satisfaction list—even among cable
providers, which the surveys also place at the bottom of industry lists.[14]

Executive pay expert Graef Crystal once called Ralph Roberts the
most overpaid businessman in America for collecting $6 million in a sin-
gle year. That was in 1996, four years before Roberts paid himself $100
million, mostly in stock-options profits, despite the company's reported
losses, which critics like Silicon Valley media investor Andy Kessler
claimed were being stage-managed to reduce Comcast's income taxes.

Morton Bahr, head of the Communications Workers of America
union, who represents some Comcast workers and is trying to sign on
more, chided Comcast's "greed" and "arrogance."[15] Bahr has lately sent

squads of red-shirted activists to the Maryland State Capitol to denounce Comcast's ties to Gov. Robert Ehrlich of Maryland (whose wife is a Comcast lobbyist) and to ask polite but pointed questions at Comcast's annual meeting. Comcast pushed back, trying to persuade its few organized workers to leave their unions.

On the eve of the AT&T deal, a coalition of Christians, from the nation's biggest Protestant denomination to the Catholic Cardinal of Comcast-wired Baltimore, denounced Comcast and the systems it acquired from AT&T as the nation's biggest purveyors of pornography. But their campaign got little play in the mainstream media, or on Comcast's own house channel, CN8.

The critics have little to show for their outrage. The company has won all of its major battles, preparing for each as if a loss would be a disaster.

As a public company, Comcast is owned mostly by big mutual funds and pension plans. But the Robertses needn't worry about shareholder complaints: Though they own less than 2 percent of the company's stock, Comcast rules, until recently, gave them absolute veto power over shareholder decisions. When AT&T investors objected to handing over absolute voting control, the Robertses settled for 35 percent—still by far the company's biggest voting bloc— but insisted on adding a provision that virtually guaranteed Brian's job for ten years, unless he votes to fire himself.

Comcast still lacks the fame, notoriety, and staying power of big-league monopolists—Microsoft or AT&T, U.S. Steel or Standard Oil—at their heights. The digital media business remains in flux more than half a century after the first coaxial cable got laid. It's far from certain whether Comcast's nation-girdling network of lines, plus the smarter and faster new video, Internet, phone, and wireless-compatible services Brian Roberts is trying to sell, will become America's nerve center—or whether changes in markets or media will make Comcast as irrelevant as the Erie Canal in an age of interstate highways and international airports.

Will Comcast take its place beside Wal-Mart and GE as a dominant national company? Will Brian Roberts and his inner circle finally sell out for billions of dollars in cash? Will they buy a big studio like Disney in a bold aggressive bid to manufacture its own programs? Will they get swamped by smarter competitors, cheaper technologies—or their own debts? Could the family somehow maintain control long enough to groom Brian's own children to run Comcast?

Whatever their fate, the extended Roberts business family enjoys a lot of influence over what you see, how you see it, and how much you pay to use the screens around your house.

Here is the story of how they gained that power.

FOUNDATIONS

In the New World

"The first thing I remember is digging up my mother's marigolds and selling them to the neighbors," Ralph Joel Roberts recalled when he was old and rich.[1] Mom was native New Yorker Sara Wahl, who nurtured a strong and cheerful self-confidence in her children. Dad had been born Ruvayn, son of Mordecai, on the Jewish fringe of the tottering Russian empire; his parents took him away from its class struggles and ethnic murder in 1904, at age 15, with a one-way passage west.[2] In New York, young Ruvayn changed his name to Robert Max Roberts, Bob for short.

Bob Roberts was a factory chemist turned pharmacy operator, patent-medicine pitchman, and property speculator, who, in the Jazz Age bubble of the 1920s, became a prominent member of the Jewish professional and business community, with immigrant roots but an increasingly American self-assertion, which was beginning to flourish in New York's Westchester suburbs on Long Island Sound. Bob moved his family north from Manhattan in 1925 to Rosehill Avenue in New Rochelle, then as now a pleasant curving street of part-stone homes and flowering trees, closer to the town's three country clubs, where Jews weren't welcome, than to its three synagogues. Rosehill ended at the most beautiful of the clubs, Wykagyl, where celebrity pros like Bobby Jones trod the newly sculpted fairways in Ralph's youth. Down North Avenue past the center of

11

town stood a club that proved more hospitable to the newcomers. Ralph and his older brother Joseph and his kid sister Shirley launched their modest sailboats at the Westchester Country and Yacht Club, with their friends from Saturday school, on the breezy summer waters.

New Rochelle was an auspicious place to raise a future media baron. It was a suburb of choice for some of the nation's most popular artists: magazine illustrator Norman Rockwell, Broadway producer George M. Cohan ("Yankee Doodle Dandy"), and later the historical novelist Edgar L. Doctorow (*Ragtime*). For them as for Bob Roberts, New Rochelle meant a couple of steps up in the world from the loud busy city where they made their money.

Bob Roberts had remade himself as an American; at the same time, he deepened, rather than cut, his young family's identification with Jewish community and causes. Sara was a local leader of Hadassah, the women's Zionist group that raised money for the future State of Israel; Bob served as financial secretary of the Hebrew National Orphan Home, founded by poor immigrants from Russian Moldavia on New York's Lower East Side before the institution, like the Robertses, moved north to Westchester County.[3]

Like the more ambitious druggists of his day, the elder Roberts invented and sold his own private-label preparations. He even indulged in celebrity marketing, paying the opera singer and World War I Victory Bond fundraiser Ernestine Schumann-Heink, one of the first recorded music stars (her greatest hits included "Silent Night" and "Auld Lang Syne"), to endorse his sore-throat spray.[4]

But the center of Bob Roberts's business world was on midtown Manhattan's Madison Avenue, where he ran his principal pharmacy in the twenty-six-story Hotel Biltmore. Flagship of a chain of resorts named for New York's fabulously wealthy Vanderbilts and served by their various railroads, the Biltmore was neither the best nor the worst of the hotel row that New York Central developed near its ornate Grand Central Station. Later gutted and rebuilt as the Manhattan fortress of the giant Bank of America (now Comcast's main bank), the Biltmore in Roberts's day served a flashy, free-spending clientele. Jazz Age writer F. Scott Fitzgerald (*The Great Gatsby*) set his first successful story at the Biltmore, where the management later expelled him for misbehavior on his honeymoon.[5] Later, in the darkest year of World War II, the Biltmore hosted the famous Zionist congress that organized the State of Israel. But by then the Robertses had vanished from the Biltmore, and from New York.

The Depression was unkind to Bob Roberts's aspirations, and his health. In cold February 1933, his heart stalled; lingering three days, he died at his home on Rosehill, attended by his doctors, the city's fire chief, and a crew of police and electric-company workers who had tried frantically to save him. Members of New York's Jewish elite filled the house in tribute; pallbearers included state Supreme Court Judge Aaron J. Levy and federal magistrate Louis Brodsky, later famous for refusing to jail anti-Hitler rioters. The service was jointly conducted by Rabbi Louis Schwefel from the Conservative Temple Beth El synagogue and Rabbi Alvin Luchs of Reform Temple Israel. The local papers listed the visiting dignitaries; the *New York Times* marked the death in a back-page paragraph. Ralph was twelve.[6]

Philadelphia

The widow Roberts spent four years trying to hold her children's world together. Sara sold the house; she sold insurance to friends and acquaintances. Ralph sold ads for a bus timetable and helped run a neighborhood dance band, while his older brother prepared for college. "After my father's death, my mother explained to her children that we didn't have all the money that we had before, and we had to live more frugally. We weren't destitute by any imagination, but we were all made aware of the fact that an economic change had taken place," Ralph said forty years later. "I didn't want to ask my mother for money, so I tried to make a few extra bucks."[7]

But in 1937 Sara liquidated what was left of her husband's estate and moved with Ralph and Shirley to Philadelphia's leafy Germantown, near her sickly sister, leaving her firstborn, Joseph Wahl Roberts, enrolled at New York's elite Columbia University. One of Philadelphia's oldest neighborhoods, Germantown has aged badly; but in the 1930s it was still a porch-lined streetcar suburb, a place with aspirations. Ralph would be an Ivy Leaguer like his brother, despite the forced penny-pinching. Acing courses at city-run Germantown High School, Ralph acted in his senior play and served as manager for a dance band that played at the city's public high schools. He won admission to the University of Pennsylvania's Wharton School of business, a practical-minded institution with liberal roots, a favorite of ambitious immigrants' sons, where he studied economics and other forms and theories of what he was already putting into practice. Faced with this new financial challenge, he sold more retail ads, this time on desk blotters he distributed to students in competition with the post office; and he arranged deliveries of milk from Abbott's Dairy in Society Hill and eggs from South Jersey's immigrant chicken farmers[8] to fraternities at the University of Pennsylvania.

He still liked music. At a dance, he met Suzanne Fleisher, bright and talkative and bleached blonde, a daughter of Philadelphia's wealthy and accomplished German-Jewish elite, which combined a broad social outlook with a profound insularity. Her father and stepfather ran a Philadelphia investment house that was a direct ancestor of Michael Milken's storied, infamous 1980s junk-bond trading firm, Drexel, Burnham, Lambert. She acted in productions of the Neighborhood Center in South Philadelphia, a settlement house that her family had helped found two generations earlier for poorer immigrants, like those who sweated for piece-rate at big factories such as the Fleisher family's giant yarn mill.

Sara Roberts died in Philadelphia in 1940, before she could see Ralph graduate. She was laid to rest beside Bob, back in New York.

Ralph would stay a Philadelphian. He took his Penn degree in 1941 and soon added a naval commission and a posting at the Philadelphia Naval Base, convenient to his blooming courtship with Suzanne. Ralph would leave work, cross South Philly, and, in his navy whites and ensign's stripes, "patiently stand at the stage door waiting for Sue to finish rehearsals," Roberts's top deputy, the late Daniel Aaron, recalled in his memoirs.[9]

The year Ralph haunted the stage door waiting on his Sue, Aaron worked down the hall at the Neighborhood Center as a youth program director, and towering future Comcast financial guru Julian A. Brodsky was slamming low-arced basketballs at the hoop in the center's low-ceilinged gym. The three men, all graduates of Philadelphia public high schools, wouldn't formally meet for twenty years—but, to Aaron, their shared history helps explain the easy if sometimes squabbling syncopation of their later long collaboration.[10]

Ralph married Suzanne the next summer. He was twenty-two, and his biggest regret was not marrying sooner, he told Ted Turner's Cable News Network half a century later.[11] Ralph had the subsequent good fortune to spend World War II fighting Nazis from an accountant's desk, one of many that helped keep track of the loud machine shops and busy dry docks of the navy yard.

Her Fortune, His Future

"The coup that Ralph pulled off was to get Sue to marry him," said one shrewd media-watcher, who has marked Roberts's progress for more years than she cares to admit. As with the Kennedys, the Roberts family roots give their reputation a certain immunity not granted to ordinary politicians—or CEOs: "If you take that kind of alliance, a marriage that sustains itself over fifty years and five kids—and pair that with a monopoly that ought to raise an awful lot of questions—the family makes it all somehow bigger than those questions."

Ralph Roberts's career was sustained by his energy and his brains, and supported by his wife's many talents; but his rise in the world was propelled by the Fleisher family's money and connections. His orphaned family, for all the early prominence it had enjoyed in New Rochelle, was nobody in particular in Philadelphia. Suzanne's brother, Robert Henry Fleisher, and the family's longtime allies in Philadelphia's Jewish legal and investment communities, would prove crucial to Ralph's career as an investor, entrepreneur, and cable pioneer; family connections even helped his son Brian land his first and only job outside his father's company many years later. Without critical and timely support from the Fleishers and the prosperous Philadelphia business and professional network in which they had moved so long and so successfully, Comcast might have languished unborn on some distant corporate star.

Birthright

Suzanne Fleisher Roberts's frenetic, dominating energy enabled her to launch successive careers as a stagestruck actress and radio producer, public television advocate and political advertising pioneer, globetrotting amateur social worker and therapist—all while serving as the consort of a rising captain of industry and raising five children, with a full social schedule on the side. Yet, in the written-for-women feature stories she still cultivates from the Philadelphia press, Suzanne has made a point of how hemmed in she felt by her own mother's litany of what a young lady wasn't supposed to do; she complains of being forced to lighten her hair as a child, of being hit and pushed around by the domineering women in her family. By contrast, she credited positive Sara Roberts with building confidence in Ralph—who in his turn has said he sought to encourage, not criticize, his own kids.[12]

For all her complaints, Suzanne's background seems designed to have bred invincible self-confidence. The Fleishers traced their origins to small-town south Germany; the family had moved west to America in the 1830s, when economic depression, combined with legal restrictions on Jewish inheritance, drove scores of ambitious immigrants to Pennsylvania and its many German communities in those years. The first Fleishers sold supplies to upstate lumbermen and miners and oilmen; a generation after their arrival, one was arrested for running the Civil War blockade to trade illegally with the Confederacy (he claimed he'd been forced to do it). Over time the extended family's concerns included a block-long yarn mill in South Philadelphia—complete with a flourishing mail-order business, championship soccer team, and a sweating army of mostly female piece-rate workers—in addition to an office building in Center City.[13]

Suzanne's mother was the former Selma Gerstley, whose family were senior partners in the city's premier Jewish stockbrokerage. When her service as a visiting Lady Bountiful to poor immigrants from Poland and Russia was checked by her lack of Yiddish, she addressed the chosen objects of her charity in her own family's ancestral German, according to an old newspaper account, which failed to note how the Yiddish-speakers responded.[14]

Suzanne's father, Alfred, went into commercial real estate; his cousin Samuel inherited the management of the Fleisher woolen business. They owned neighboring mansions in Philadelphia's historic Fairmount district, in the shadow of the Greek-temple hilltop complex of the Philadelphia Museum of Art. Samuel's Green Street brownstone, at nineteen thousand square feet as big as a whole block of his mill workers' rowhomes, was recently renovated as the palazzo of one of Pennsylvania's most powerful politicians, State Senator Vincent Fumo.

Samuel served on the city welfare and park boards, and built the eclectic Fleisher Art Memorial and free school in a failed South Philadelphia Protestant church mission and parish house; it is today the extended Fleisher family's best-known monument. Similarly, Alfred headed the trustees of the grim stone Eastern State Penitentiary, and claimed in a lecture at Philadelphia's Rodeph Shalom Synagogue, the grand temple of German-American liberal Judaism in the city, to have been bribed and threatened by gangsters, whom he mysteriously declined to name, even to the police.

Alfred died, on or about Christmas Day, 1928, after a wasting illness and blood transfusions at the city's old Jewish Hospital. As with the Roberts tragedy four years later, news accounts of both paternal deaths stress the modern, failed efforts to revive the patient, while passing over the actual cause of their troublingly premature departures. Suzanne was seven when Alfred Fleisher died; for years, she has said, he visited her in dreams, asking, What have you done for others?[15]

Alfred Fleisher left his children an inheritance that was eventually worth over one million scarce Depression dollars, plus art, stock, and such property as the family estate in suburban Wyncote, following a long and all-too-public trip through the state courts to unwind its complex tax shelters.

The Family Way

Married young, fatherless Suzanne and orphaned Ralph put off making babies for their first eight years together; only after Ralph was more or less settled in business did children arrive, in two clusters, at the beginning and then at the end of the fertile 1950s. The family settled at first in Wyncote, in a house named "Treetops," furnished in a Colonial style from Suzanne's own antiquing, as if to extend her family's long claim on the country.

As a young wife, Suzanne continued to act in little theaters like the Bucks County Playhouse. Her proudest roles over the years included outspoken Queen Eleanor, in *The Lion in Winter*, and, in *Lysistrata*, the energetic organizer of an antiwar sex strike in ancient Greece. She courted publicity, and gave interviewers multipage press releases telling how she had studied acting alongside the director Sidney Lumet, and once shared a stage with Southern grand-dame Talullah Bankhead. For Philadelphia radio stations, Suzanne wrote and performed short dramas, featuring Eleanor Roosevelt or her own widowed mother as characters.

Even as a 1950s suburban mom out in Wyncote, Suzanne found two very downtown outlets for her Fleisher energy and sense of noblesse oblige: Philadelphia's young Democratic Party reform campaign, led by Joseph Clark and Richardson Dilworth, who sought to expel the ancient corrupt Republican machine from City Hall; and the new mass communication medium—television—which she saw, at first, as a vehicle for education and government service. After he succeeded Clark in 1955, Mayor Dilworth tapped the prolific matron for the founding board of the new city-owned TV station, WHYY, which was charged with creating educational and cultural programs for young viewers.

Suzanne continued another family tradition by inviting the attention of the Philadelphia newspapers to her extracurricular activities. Philadelphia's own Benjamin Franklin had counseled readers in search of earthly immortality to "do the things worth writing of / Or write the things worth reading"; she would attempt both. Her father's and uncle's front-page crusades for prison and welfare reforms presaged Suzanne's own early television programs on voting, civics, and education; as her family grew, she began appearing in women's-page stories about fashionable, accomplished matrons who were successfully able to combine career, family, and home. One such profile, after listing Suzanne's civic and matronly duties, took pains to stress her femininity, as demonstrated by the ruffles of her dress.

But for Ralph, of course, it wasn't enough to marry a Fleisher—even such a paragon of artistry, service, and femininity as Suzanne. He was determined to follow his father's example and his mother's advice, and make his own way in the world.

Bent Putters

Ralph Roberts's first step into his natural vocation of salesman–investor was an underhanded failure, which Ralph has characteristically turned into one of his favorite stories, told with varying emphasis on his sharp business practices and his naiveté, depending on the audience.

"My first business when I got out of the Navy was making golf clubs," he told a Media Institute awards banquet in the fall of 2002, a time when he was collecting "lifetime achievement" awards from entertainment-industry organizations in a sort of victory lap. "Since World War II had just ended and most of the manufacturing companies were still tooled up to build war materials, no one was producing golf clubs."

Before settling on golf clubs, he and a naval base engineer had put together a little company, Altair Corporation, and experimented with scented inks and labor-saving machinery for the soda-pop bottling plants that in those days could still be found in every Philadelphia neighborhood. The putter seemed more saleable. It certainly sounded more scientific: It was a "centric" putter, with the shaft stuck in the middle of the club head. But how to promote it? Like his father, the sometime patent-medicine salesman, Ralph Roberts chose celebrity sponsorship, and did it on the cheap: During a visit by comic actor Bob Hope, then the nation's most popular movie star, to North Philadelphia's old Town Hall theater, Roberts convinced a newspaper-photographer friend to smuggle him backstage to meet Hope, who had toured widely to boost the morale of U.S. troops during the war.

Roberts approached the star, putter in hand. "'Mr. Hope,'" Roberts later recounted, "'I'm a veteran. I've started making golf clubs. You're a world-famous golfer. Would you mind if we took a picture of you holding this putter?'" He said, "Sure, kid," and he took the club, examined it, took some practice strokes and we got wonderful pictures."[16] Next, "We come out with the Bob Hope Centric Putter, and forget to ask if we can promote it like that."[17]

Roberts printed a brochure featuring Hope with his club, urging golfers to "Get on Board." The boy who wasn't welcome at the country clubs back home was now touring the East Coast's great courses, demonstrating the putters to golf pros personally. Altair was overwhelmed with orders, according to Roberts. But, in rushing their product to market, his partner, the naval base engineer, had ordered a batch of shafts from untreated metal. They bent, Ralph has said at various times, like "a curlicue," "spaghetti," "a pretzel."

And that was the end of Roberts's first adult venture. He and his colleagues had neither the means nor any plan to replace his defective products. As he later recounted, "When you know you have a bad product out there, you close up and run for the hills."[18] Roberts's marketing strategy hadn't failed; the problem lay with execution. He couldn't afford to leave that to others.

SALESMAN

The Ad Man

After his first celebrity product collapsed in a comic sham worthy of a Warner Brothers cartoon, Ralph Roberts picked himself up, weighed and applied his strengths, and resumed his rise in business. He tried out as an ad man, left for a management post, bought control of his employer, diversified the company, and parlayed his ownership stake and his wife's connections into a stint as a diversified venture capitalist, before the fateful made-in-Philadelphia investment that made him the unlikely cable TV czar of Tupelo, Mississippi, and points far beyond.

Advertising was a natural step. His success at selling even a defective product only confirmed the powers of persuasion and promotion he had honed in his schoolboy youth. Roberts wasn't the only one who believed in his golden touch: He landed his next job by cold-calling, with no apparent prior connections or help from his in-laws.

In 1950 Roberts joined Aitkin Kynett, a Center City ad agency. He later said it was the one listed first in the phone book, so he called there first. Philadelphia in those days was a center of the ad trade. It was never a literary center to approach Boston, New York, or even Chicago; but Philadelphia's founders had encouraged popular literacy even as they distrusted elite education, and for a long time the city was a flourishing workshop of mass-market publishing, home to leading printers of

Bibles and farm and medical journals, as well as giant national publications like the Curtis family's *Saturday Evening Post* and Walter Annenberg's *TV Guide*, each in its day America's most popular magazine. The
publishers attracted a row of advertising firms, led by pioneering N. W.
Ayer, which for a time vied with New York's Madison Avenue and the big
Chicago agencies for national accounts.

Aitkin assigned Roberts to a New York account, Muzak, which sold
and serviced the unthreatening recorded music once heard in elevators,
bus stations, and dentists' waiting rooms. It didn't seem to bother him
that the product became, like Spam, a byword for mediocrity and even
annoyance. Roberts developed a long affinity for Muzak; it was a cheap-
to-run cash business with few fixed costs and fewer competitors. Ralph
would hold a string of Muzak franchises, run by his brother and partner
Joseph, long into his Comcast career.

The arrangement had another benefit: It put Roberts in close con-
tact with Muzak's owner, William Burnett Benton, an American original
who straddled the worlds of business and government and achieved
some prominence in both, in his time. In fact Benton's career held many
lessons, some of them cautionary, for a young man who would later
make his living in a government-regulated industry.[1]

The Big-Government Businessman

Benton was a creator of the professional opinion-making machinery
that took control of American public life through the new media in the
second half of the twentieth century. A Chicago ad man, he invented the
soap opera and the consumer product survey. Later, as an early New
York venture capitalist, Benton owned Encyclopaedia Britannica as well
as Muzak. As a New Dealer, he helped organize the United Nations; as a
Cold Warrior he set up the Voice of America; and as a loyal Democrat
who bought ink by the barrel and tank, he hired party standard-bearers
like Adlai Stevenson and Hubert Humphrey to write well-paid articles
for his publications. President Harry Truman rewarded Benton for his
services by supporting his bid for a U.S. Senate seat. In a 1948 special
election, Benton beat Republican Prescott Bush—father of George H. W.
and grandfather of George W., future presidents—to represent Connecticut in Washington, D.C.[2]

Benton was a liberal businessman who prefigured the "New Democrats" of the Bill Clinton era. Like them, he saw no contradiction between
permissive social views, activist government, and entrepreneurial capital-

ism—especially if that meant the government helping liberal Democratic capitalists build new kinds of companies that prospered in times of social change. Some Americans imagine all businesspeople are Republicans; in fact, industry leaders in insurgent businesses like advertising, "independent" cinema, cable TV, and even computer software have included a surprising number of Democrats, who have often sought government help in challenging what they saw as unfair advantages for entrenched, conservative, Republican industrial elites. When he hired Roberts, Benton was at the pinnacle of his influence—and in the fight of his political life: His senate seat was threatened by a nasty contest with bullying anti-Communist Sen. Joe McCarthy, of Wisconsin, who was backing Benton's Republican opponent in his Connecticut reelection campaign.

Benton's politics were hardly mainstream corporate, especially for the 1950s; but they were nothing strange to Roberts. Like others in her family, Ralph's wife Suzanne had been working to help Philadelphia's Democratic reformers against the city's century-old, entrenched, corrupt Republican machine. Suzanne's sister-in-law, Mrs. Leon Sunstein Jr., even became an officer of the Americans for Democratic Action, which combined support for the "progressive" tradition of Franklin Roosevelt's big-government New Deal with a vigorous, defensive anti-Communism.[3]

Impressed by Roberts's political ties, Benton diverted the young ad man from selling canned music to scribbling good lines for his Senate campaign. The race had national implications; control of the Senate, and Franklin Roosevelt's New Deal legacy, turned on a handful of seats. One day Benton called Roberts "to say that he'd be in St. Louis tomorrow and needed one snappy sentence to give the reporters. Then I had 20 minutes to come up with the sentence," he recounted.[4]

That was too much. Roberts didn't like copywriting, he wasn't passionate about politics, and, after writing the snappy line, which he says the candidate liked but he himself long ago forgot, Roberts determined to quit his job with Benton, who would go on to lose the election. So Roberts said good-bye to his brief career as a political image-maker, though not, as it turned out, to Benton, whose liberal ideas, given an afterlife by his business fortune, would dog Roberts's company and other television enterprises.

The Benton Foundation, charged by its late founder with monitoring such things as "the relationships of Congress to broadcasting, as well as the impact of the new media on political campaigns," has decried Comcast's lobbying power and demanded more competition and free public-service programming from the company. Benton himself, however, had used government contacts and personal fortune to serve his own ends; as the *New York Review of Books* wrote in 1970, Benton's career "proved that in America there still isn't much that money can't buy."[5] If, as the foundation claimed, cable operators used political influence to clear their business road, Ralph had a master teacher in William Burnett Benton. Of course, it wasn't Benton, but the Bush family, whose ambitions he briefly thwarted, who would later show what might be achieved by the wedding of corporate and political clout with effective media strategy.

Ralph in Charge

A month after quitting Benton in 1952, Roberts took a job as "assistant to the president" for promotion and ads at Pioneer Belt & Suspender Company and its factory in dowdy Darby, a fading suburb west of Philadelphia. Pioneer's customers included the U. S. Army, which bought more than half a million suspenders from the company during that year. Ralph has said he found the job through a headhunter, but he could as easily have asked around among his wife's circle of friends: Pioneer's president was Leo Heimerdinger Jr., another old-line Philadelphia manufacturer and Jewish philanthropist.[6] Joining Pioneer, Ralph ended his New York commute and reduced his absences from his wife and new daughter, Catherine, and set about proving himself. Heimerdinger had succeeded his own father to the company's top job, but he had no son of his own, and his health was poor. Ralph Roberts was anointed; after a year he was promoted to second-in-command and heir-apparent.

Pioneer belts were sold by thousands of retailers, and Ralph threw himself with gusto into the work of building sales. In 1953, Ralph introduced product innovations such as cufflinks and tie clasps shaped like antique cars and pistols, based on what he said was market research about men's favorite hobbies. By 1955, Ralph had bought control of Pioneer and succeeded Heimerdinger as president; the older man remained chairman, but Roberts was joined on the board by his brother-in-law, Robert H. Fleisher, and his wife's stepfather, stockbroker Leon Sunstein Sr. Roberts put up $35,000 of his own money; Fleisher put up more. They borrowed still more through loan officer Jack McDowell, who handled the Pioneer account for the proudest of the city's old-line lenders, Philadelphia National Bank.

Here was yet another of the fortuitous alliances that would help Roberts build a national company: McDowell would eventually rise to PNB's chairmanship and put Roberts on PNB's board of directors; PNB would be one of the first in a string of banks, insurers, and other professional investors that would bet on Comcast as one of the more solid properties in the shaky, cash-starved cable industry; the bank helped supply the piles of cash that funded Comcast's relentless acquisitions.

Pioneer called itself the second-biggest belt maker in America. Roberts went to work on the company's image, putting together fancy store displays and flashy promotions during New York's yearly fall Fashion Week. Pioneer even provided a stage for Suzanne, who starred in company-promotion shows and gloried in the Manhattan spotlight. Ralph renamed the company Pioneer Industries and nurtured upscale men's accessories like pricey Mark II cologne—"The Mark of a Man"—sold in a phony-gilt flask like expensive whiskey. Roberts gave every sign of having enjoyed selling men's accessories; even after he went into the cable business, Comcast veterans recalled Ralph's fussing over new packaging shapes and lettering fonts, and spraying musky new scents on pesky visitors who pressed him on matters of urgent business.

In charge of Pioneer, Ralph soon ran afoul of its larger competitors. He personally attended to such details as the design of the cologne bottles, which under Ralph's hand came to resemble those of Fabergé, the better-known, pricier, New York–based scent maker. At least, that's what Fabergé thought, and it threatened to sue Roberts and Pioneer for copying its trademark packaging. Roberts and his attorneys at the firm of Wolf, Block, Schorr & Solis-Cohen—yet another of the premier Philadelphia German-Jewish businesses that welcomed Ralph and his Fleisher connections—did what Ralph's old boss, marketing trailblazer William Benton, might have done for a similarly pressed client: They ordered a consumer product survey, then told Fabergé they had scientifically proven consumers weren't confusing the bottles. Apparently, in those days, before dueling expert witnesses, such a survey carried weight. Ralph then carried the day by suggesting Fabergé help him design new bottles.[7]

Even if you still suspected him of pirating your product, how could you stay mad at a guy like that? Fabergé decided not to sue.

Roberts also faced legal action by Hickock, Pioneer's larger rival, which, like Pioneer, threatened a court complaint over Roberts's alleged use of Hickock designs. Here again, the Wolf connection proved useful: When Hickock's lawyer, a Talmudic scholar, began making learned citations from arcane Hebrew ethical treatises in open court, Roberts's lawyer, Abraham L. Freedman (whose wife Jane was Suzanne's stepsister), "matched his adversary quote for quote and phrase for phrase," Roberts later recounted. Hickock ended its complaint.

But Roberts wasn't about to make masculine enhancement his life. Hickock kept up the pressure; finally, tired of the competition, Roberts agreed to sell Pioneer to his more successful rivals in 1961. Roberts has gleefully credited his decision to negotiate the sale to his own foresight that the beltless polyester slacks of the 1960s would come to devastate Pioneer's business, but he also took pride in unloading Pioneer for several times what he'd paid Heimerdinger six years earlier. Roberts didn't sell everything; he kept Pioneer's line of colognes.

The Road Not Taken

Ralph Roberts had the skills and the capital to have spent the rest of his long working life trading businesses and building the kind of ever-changing, mostly local conglomerates that came into style as he entered the prime of his business life in the 1960s. And Roberts did indeed begin, like his Philadelphia contemporaries Ray Perelman (father of Revlon chairman Ron Perelman), Robert Fox, and Warren "Pete" Musser, the business of assembling a loose, disproportionately local industrial conglomerate, on the cheap. Like them, Roberts was an early venture capitalist. But unlike Perelman and Fox, who in a lifetime of mostly below-the-radar-screen deals built fortunes that enabled them to endow local schools in their name, or Musser, who briefly achieved national prominence and paper billionairedom during the Internet bubble, only to lose it in the collapse, Roberts eventually moved beyond the magic of mere deal flow and came slowly to concentrate on dominating a single industry.

He had grown tired of making money the hard way. His dad had struggled in the drug trade; his in-laws had watched the woolens business shrivel; he had just spent a decade battling in the fickle, low-margin men's clothing trade. He was tired of old industries; in a country growing this fast, he reasoned there must be an easier way to make it big. Roberts, in short, "was looking for a new business that would not be as competitive as the one he had just sold," Aaron recounted in his memoirs.[8]

Roberts reorganized his holdings as the grandiose "International Equity Corporation," and went to work looking for deals. He had collected what he later called a "sizable cash stake" from the sale of the Pioneer belt and suspender business. International Equity soon assembled a string of unrelated businesses—cologne, punch-card computers, even the franchise of his old employer, Muzak, that had rights to the towns around the new Disney World in Orlando, Florida—when Roberts stepped out into Center City Philadelphia and found himself accosted by Musser, a casual acquaintance and small-time investor. Musser then saw in his own cable TV system in strife-torn northern Mississippi a liability in need of three things he lacked: ready cash, a capacity for quick action, and a long-term view.

PIONEERS

3

Live from Philadelphia

The crowd stood three-deep outside the little appliance store in the hard-coal borough of Mahanoy City, Pennsylvania, watching Ralph Roberts's wife perform. There she was, tall and curvy, in black and white, on the little glass screen of the big boxy set, urging Philadelphians to get out the vote and demand better treatment by the politicians who worked for them. Philadelphia was and remains a hundred hilly miles away. Yet people watched, because out-of-town programming was all there was, and because, as Ralph and Brian Roberts are fond of saying, people love television. So they watched Mrs. Ralph in Mahanoy City, before young Brian was born, while two nearby sets beamed quiz shows, piano players, weather instruments, comedy routines, men reading Korean War news in radio staccato, and all the other earnest programs of primitive TV: three channels, and each running side-by-side on a separate set, reminding the viewers they had a choice.

It was 1951, and Suzanne Fleisher Roberts, Philadelphia Democratic Party media guru, author of the soon-to-be-published primer *The Candidate and Television*,[1] and new mother, was creating the modern American political campaign over the screens' chatter. She had no reason to suspect the programs she produced in Philadelphia could be viewed way up in Mahanoy City. The anthracite towns, still flush in those days with the unaccustomed wealth of the war years, were outside broadcast TV viewing range. They still are. Though Mahanoy City is not that far up the twisting Schuylkill from Center City Philadelphia, it sits in a gulch that might as well be on the other side of the moon for broadcasting purposes, surrounded by long stony oak-covered hills that bounce the urban signals harmlessly into space. It's that way, more or less, from the Delaware River to Lake Erie: Pennsylvania could be bigger than Texas, if you hammered it flat. It's as if God were giving the people of the valleys every chance to avoid big-city broadcasters.[2]

But for practical and enterprising businessmen in places like Mahanoy City, this problem as tall as the hills was also an opportunity as wide as America. By 1951, the appliance dealer and the chief of police were competing to wire this particular town to hilltop towers, and others were busy in burgs to the north and west, connecting their citizens to the strange new world of Philadelphia broadcast television so men like them could sell more TVs, and collect from every TV family, every month, for the privilege.

For the Roberts family—and their allies in marriage and finance, the Fleishers—their first cable TV investment was still more than a decade in the future and a thousand miles away. But the amplification of Suzanne's broadcast performance to a cable audience in the Pennsylvania hills was a flickering portent. Who could know that her family's involvement in the industry would ensure that, half a century later, she would still be performing on cable TV?

"The Greatest Thing Since Stealing"

"It's the greatest thing since stealing," Comcast cofounder Daniel Aaron wrote in his memoirs.[3] "These guys put an antenna on the top of a hill, run wires through the trees to a central location, and then hook up individual homes to receive a clear television signal. They charge their clients $100 to hook up, and then use the $100 to wire up the next street. There is a three-buck-a-month service charge. They don't pay taxes. They keep it all."

Many years later, when Comcast had become a big, prominent company, Ralph replaced the criminal comparison with a folksy note for a national publication: "I was never, never nervous about buying a cable system," he said. "You have recurring billing, reasonable rate increases, you keep your costs down and it's like chicken in a grocery store. It's very nice."[4]

But early cable was better than stealing. The pioneers didn't even pay for programming: They just collected broadcast signals from the air up the ridge, high in the trees, then sold it down the hill, sometimes in competition with another local entrepreneur who had strung his own wires from a hilltop antenna down through the neighborhood in search of customers. All this meant more business for Philadelphia, then a media-hardware workshop. Local factories made radios, TVs, record players, and vehicle electronics, even the first commercial computers (UNIVAC). The cable men bought TV equipment from Philadelphia's wholesalers and manufacturers—and advice from the city's lawyers and other professionals.

Among the most avid students was young Julian Brodsky, newly graduated from the University of Pennsylvania's Wharton School. Brodsky had labored on the political left, working on the staff of the International Ladies' Garment Workers Union. He had started his career as the lone Jew among Irish Catholic accountants at William C. Howe & Company, which included among its clients half a dozen upstate cable operators whose names reflected the ethnic goulash of the coal regions: Gans, Reinhold, Walsonowich. As monthly cable sales jumped and the operators' tax obligations followed, Howe sent Brodsky north to reorganize the cable operations as personal corporations, which paid a smaller government bite. Brodsky, who had already shed his youthful socialism, liked these independent, risk-taking clients, though he soon left Howe to gain wider experience.[5]

He applied to Philadelphia's Adler, Faunce & Leonard because "they were the first firm listed in the phone book"—the same way Roberts said he chose his ad agency employer a few years before. Adler's clients included Pioneer Industries; when Roberts sold the company's belt works, Brodsky was sent to his office to talk about what to do with the money and how to keep it from Uncle Sam. On repeat calls, Brodsky found a new vocation: to hang around Ralph Roberts for the rest of his working life.

"Ralph is the most extraordinary human being I have ever met," Brodsky said, four decades later. "Talking to him is like having a perpetual subscription to the *New Yorker*. He can talk on any subject. And he can listen. He hears everything. He is very strategic. Always thinking. He stands in the background, and hears everyone, and then he makes things happen."

Brodsky, six-foot-two and as loud as Roberts was modulated, eventually served thirty-seven years as Comcast's official financial advisor. He prided himself on "creative" fundraising, but also on a "conservative" business and accounting style that kept company employees out of prison amid high temptations. He designed the complex legal structure of hundreds of state-tax-retardant Delaware subsidiaries that helps make Comcast's annual reports as thick as a small city's phone book. Brodsky had liked cable, but not enough, in the abstract, to throw over his old career for it. But now he wanted to do cable—for Ralph Roberts. "I had thought the highest calling was to be an auditor," Brodsky said. "But I loved Ralph and cable too much to resist them."

Made in Pennsylvania

Half a dozen towns from Arkansas to Oregon dispute the title of being the birthplace of cable TV. But Pennsylvania, with its mountain boroughs and its ready access to Philadelphia equipment, professionals, and cash, was the infant industry's clear center.[6] From the beginning, even the most independent cable pioneers saw advantages to working together. From upstate boroughs like Lansford and Mahanoy City, they gathered in 1951 in Pottsville at the Necho Allen Hotel[7] to form what's now the National Cable Television Association, to cope with, fight, and exploit the arcane cable TV rules already issuing from the Federal Communications Commission in far-off Washington, D.C. The pioneers called their nascent business "community access television"; only later, when communities had less to say about who controlled their wires and chose their programs, was it popularly known as cable TV.

Sometimes the FCC gave the pioneers an unexpected gift, as when, in 1948, the bureaucrats placed a four-year moratorium on the new broadcast stations that had been spreading into less populated areas. That was a shot of pure caffeine for the cable guys, forcing places that wanted TV and didn't already have it to get wired or go without. And sometimes the government took with one hand what it gave with the other. A federal judge in 1957 ruled that the federal communications tax didn't apply to cable and ordered the government to return $35 million. But the next year another judge ruled installation charges weren't tax-protected capital expenditures, and a Senate committee pressed the reluctant FCC to step up its regulation of the free-ranging cable operators.

In Philadelphia, publisher Walter Annenberg's new television guide, soon to replace the cross-town *Saturday Evening Post* as the nation's most popular weekly, wrote some of the first news accounts of the new cable companies. Annenberg also owned the *Philadelphia Inquirer*, which, along with the city's rival *Bulletin*, would soon be chasing cable franchises of its own—once the pioneering years were almost over and the business entered its bargain-hunting, deal-making, empire-building stage. The stage that brought Roberts into the fray, too.

Places like Mahanoy City were already proving that people would watch almost anything that came on the screen. It didn't even have to be violent or sexy, which wasn't allowed, or feature celebrities, who weren't much interested, to draw viewers. TV was still a novelty: radio with pictures, movies in your home.

Wired

The Mahanoy City cable pioneer who shortened his jaw-cracking name from the original Slavic to John Walson was a Pennsylvania Power & Light Company lineman who moonlighted as an appliance retailer. He read a book called *Radio Physics* and taught himself to run a wire up the mountain to bring the TVs in his store to life. This helped him sell the new TVs rolling out of the Philco factory in Philadelphia and the RCA plant across the Delaware River in Camden, which retailed for around $500—about as much as a good used car in those days.[8] Walson improved service with a reliable double (coaxial) line capable of resisting the rain and delivering more than one channel; to demonstrate, he set up three TVs in his window, each tuned to the different Philadelphia affiliates of the emergent national networks.

Men like Walson wore multiple hats, serving as president, bill collector, lineman, maintenance man, engineer. In those early years there was even competition—in Walson's case, from Mahanoy City's police chief, who strung his own competing wires, just as Edison and Westinghouse and their imitators had built competing power generating and distribution systems in big cities at the turn of the previous century. Competition and the low cost of getting into the business kept prices low. In Mahanoy City at the beginning of the 1950s, Chief A. P. McGlaughlin and TV salesman Walson each charged $2 a month for cable service, compared to $3 dollars in less fortunate one-cable towns nearby; the service was even free, if you agreed to locate the neighborhood cable box in your basement and let the cable guys in whenever they needed to add new customers, cut someone off, or figure out why the block's TVs had unaccountably gone snowy.[9]

John Walson's company, Service Electric, split among his children, still provides cable TV to Mahanoy City and larger towns nearby. Monthly rates, though they've risen lately, are still more than 25 percent cheaper than what Roberts's company, Comcast, charges in the bigger towns further down the Schuylkill. But Service Electric is less than 1 percent of Comcast's size. It takes more than a lineman's skills or an appliance salesman's showmanship to build a really imperial monopoly.

The Bare-Chested Antenna Salesman

The first big cable fortune wasn't made by the upstate pioneers like Walson or the big-city moguls like Annenberg, but by the guy who sold them their equipment. Milton Jerrold Shapp was an East Coast forerunner of the computer and chip salesmen who made Silicon Valley the center of the computer business two decades after him. If he'd hooked up with struggling Univac Corporation rather than building his own antenna firm, it's possible Philadelphia wouldn't have spoiled its early lead but might instead have fulfilled its thwarted destiny to become a great computer and software center. As it was, Shapp's company, Jerrold Electronics, grew over the years from a South Philadelphia machine shop, to a North Philadelphia factory, to a sprawling facility in suburban Horsham, where it formed the core of General Instruments Corporation; today it produces digital TV equipment as an arm of the vast Motorola electronics group.[10]

Like Roberts's old boss, Bill Benton, Shapp had political as well as business ambitions. In time he became Pennsylvania's first Jewish governor, a liberal who challenged and infuriated the state's powerful insurance companies and other established interests even as he was accused of cutting deals to favor his supporters. Shapp is remembered little and unkindly in today's Pennsylvania, even among fellow Democrats; he warred incessantly with party leaders over mutual accusations of corruption and incompetence, ushering in twenty years of conservative, largely Republican, rule.

But Shapp's business career was remarkable. The son of a Cleveland hardware salesman and a women's suffragette, Shapp trained as an electrical engineer and served as a captain on that toughest of engineering training grounds, the Army Signal Corps in Europe during World War II. He sold commercial electronics after his discharge; when he couldn't find what his customers wanted from existing manufacturers, he started Jerrold Electronics Corporation in 1948 to build and market a powerful new antenna suitable for pulling in broadcast signals from greater distances. Shapp's manufacturing venture proved far more successful than ex-Ensign Roberts's contemporary golf clubs.

From single-TV signal boosters, Shapp went on to "master antennas" that could shoot multiple channels to multiple customers in TV showrooms or apartment buildings. He also began investing in and financing his clients. The new antennas and Shapp's support were picked up by cable pioneers like Robert Tarlton of upstate Lansford, whose Jerrold-backed systems brought cable to Wall Street's attention through favorable articles in the financial press.

Certain attractive provisions of the federal tax code, which allowed cable operators to write off their set-up costs over a three- to five-year period, drew in investors who didn't know Milton Shapp from TV-host Uncle Miltie. Those breaks were worth extra in the postwar years, when top tax rates were much higher than today's. The end of each write-off gave operators an incentive to sell their systems and start new ones; buyers could improve older systems and start the merry tax breaks all over again. This legal tax avoidance set the tone for the industry. It favored freewheeling dealmakers over small-town chamber of commerce types, quick-buck financiers and quick-and-dirty technicians over old-school engineers and media visionaries. For entrepreneurs like Shapp, the money was in setting up systems quickly, turning them over, and starting or improving new ones. Basic maintenance problems needed to be solved; custom programming could wait. In 1956, Goldman Sachs and a group of New York venture capitalists raised $200,000 to hire Tarlton to wire the old lumber center of Williamsport, the biggest town in Pennsylvania without broadcast TV.[11] And the empire-builders began to talk of stringing national systems together, bouncing signals through bigger and bigger antennas.

Some of the first cable users were the clubby people whose renamed fictional doubles populated local-boy novelist John O'Hara's contemporary bestsellers—the eye doctor, the car dealer, the dentist on the outskirts of the coal towns—but, as cable caught on, Shapp and his more successful customers focused on the working-class rowhouse neighborhoods that adjoined the little Main Streets. They were cheaper to wire, for one thing. Jerrold became, for a time, the nation's biggest cable operator, running franchises from Dubuque, Iowa, where citizens actually chose companies in an election, to Memphis, Tennessee. Some early programs were beyond primitive. One channel relayed a wire-service news ticker; another showed a fishbowl and played background music; another featured a thermometer and wind gauge. It became a standard cable joke years later that these were the precursors of CNN, MTV and the Weather Channel, respectively.

It wasn't all work and no play. In his Philadelphia factory, Shapp extended his mother's liberal principles, hiring Ivy League socialist supporters of former Progressive Party presidential hopeful Henry A. Wallace to some top management jobs. He made a point of employing

African Americans and promoting female managers; he also showed he could mix with the masses, sort of, when he danced bare-chested at Jerrold's drunken Christmas parties. Shapp came to terms with Philadelphia's powerful electrical workers' unions, built bridges to the Democratic labor leadership and, later, formed ties to President John F. Kennedy's youthful administration. Like Senator Benton, Shapp wasn't shy about his unorthodox politics; he advertised it. He even claimed credit for encouraging Robert Kennedy to start the Peace Corps. Eventually, in 1966, Shapp sold his company, focused full-time on his political career, and departed cable history.

But men who learned their trade at Shapp's company continued to shape the industry. There was Daniel Aaron, the Philadelphia *Bulletin* reporter so taken by Shapp's charisma that he quit journalism to work for Jerrold after writing a profile in 1955; he eventually brought Roberts into the cable business and became Comcast's energetic, supremely commonsensical operations chief. Finally, in the late 1960s, Shapp's successor hired a whiz-kid consultant, an aggressive Yale graduate with Pennsylvania roots, to clean up the mess Shapp's socialists had left in the office when he departed on his political crusade. The consultant told his biographer that Shapp's financial managers had "pumped the turkey" and left a dangerous situation that could only be rectified by a thorough reorganization; but what else do consultants ever say?[12]

This particular consultant, John Malone, cleaned house at Jerrold, then headed west to Denver to build the TCI cable empire, and later Liberty Media, a grab-bag of TV programming; his contemporaries still consider him the most brilliant of the industry's many visionaries, his leadership marred only by the aggressive bluntness that made him a poster child for cable-regulation advocates and populist politicians like Vice President Albert Gore. But cable would need a gentleman, not just a genius.

The Big Fish

Aaron quit Jerrold in 1962 to set up a cable brokerage, buying and selling cable systems to tax-adept investors. An early client was Philadelphia's own Warren "Pete" Musser, who, in strange and distant Tupelo, Mississippi, singer Elvis Presley's hometown, controlled a twelve-hundred-subscriber system, which he wanted to liquidate so he could pump his money into the rising stock market. "Pete," as Brodsky later put it, "was suffering one of his periodic liquidity crises," and chasing the next big thing on Wall Street. Conglomerates were in, so were electronics companies—and that's where Musser wanted to move his money once he got it back to Philadelphia.[13]

But first he needed a buyer. Mississippi in the early 1960s roiled with racial and political strife that made it as unattractive to many investors as an unstable third-world country. Tupelo needed someone with a sense of adventure, a bargain-hunter willing to come down, look the place over, and make the friends he would need. So did the three as-yet-unbuilt cable systems Musser had purchased when he bought Tupelo. Musser hired Aaron, the cable broker, to find this ideal buyer, or at least someone with big ideas and ready cash. As Aaron told the story, he and Musser were walking down Philadelphia's crowded Chestnut Street[14] one day when Musser smiled and announced, "Here comes our 'fish.'" It was Ralph Roberts, dapper in a fine raincoat, "flush with cash and ambition" from his suspender-company sale.

Musser knew Roberts from Philadelphia's Young Presidents' Organization of youthful businessmen. Shapp had addressed the group and told them his vision of cable's vast promise; it wasn't that unusual for wealthy Philadelphians of that period to consider an investment in a cable franchise. So Roberts was ready at least to hear Musser's pitch.

It was Aaron's job to explain the business to Ralph, Ralph's brother and adviser Joseph, and brother-in-law and co-investor Robert Henry Fleisher. The group gathered at the home Roberts had built for his family in Elkins Park, a comfortable suburb north of the city; the family chatted for hours with their visitor, now talking familiarly about sailing and Shore towns on Long Beach Island, now dragging the scariest details of the Mississippi franchise from their visitor. For all the obstacles, they liked the sound of the customer cash flow, and the potential for growth. The biggest problem, Aaron later joked, was getting the late-rising Ralph Roberts back together with early-bird Musser to negotiate the final price.[15]

In 1963, Roberts's International Equity bought the Tupelo franchise for half a million dollars, and Roberts hired Aaron to run it. The Roberts and Fleisher families put up 51 percent, the only personal money they ever invested in Comcast, Ralph boasted years later.[16] The rest was raised by the Philadelphia friends and longtime business associates of the Fleishers: Fred Wolf and a handful of his partners at Wolf, Block, Schorr & Solis-Cohen, the city's premier Jewish law firm; Ray Perelman, the Philadelphia investor and father of New York corporate raider and serial stock diluter Ron Perelman; and the Sunsteins, co-owners of Philadelphia's leading Jewish brokerage and Roberts's step-in-laws.

It wasn't yet called Comcast; the company would be International's American Cable Systems division until the end of the 1960s. Ralph had to grant board seats to several of his investors. Yet none of them had an equal say. Ralph had one key demand: that he control a majority of the voting stock. As his share of actual ownership steadily diluted with the mergers of the next four decades, Ralph and his son retained a stranglehold over the company through their "supermajority" voting rights, until Brian Roberts's crowning deal with AT&T.

Soon afterward the Tupelo deal closed, and Brodsky quit his accounting firm; he told Roberts, "You're not doing this without me." Late in 1963 he brought his own folding chair and table to Ralph's modest office on the eleventh floor of the stark cement tower at Monument Road and City Line Avenue at the western edge of town (like too many Philadelphia towers, it still bears the logo of a vanished bank, Germantown Savings).

Ralph Roberts was in charge; he would take the lead in negotiating the expansion of the cable business, while overseeing his men's toiletries line, along with future investment opportunities, including his brother's personal favorite, cash-rich Muzak. Aaron would run the cable business, which meant a lot of travel; Brodsky would work the books, which to a veteran auditor meant lots of time in the field kicking tires. Joe Roberts, Bob Fleisher, and the guys at Wolf, Block and Gerstley, Sunstein would give their opinions, needed or not; a small staff of experienced office workers, all women, kept the wheels roll ng. With the company set up like that, Ralph almost guaranteed conflict among his strong-minded lieutenants; he'd sit back and listen to all sides, then render his decision smoothly. The team was set.

DEEP SOUTH

"Boom, There's Explosions"

"It was a very strange time," Julian Brodsky recalled, "for a couple of Jewish guys from the North to be wandering Mississippi." That summer of 1963, as the Roberts team headed cross-country to check on their new Tupelo cable assets, committed liberal Northerners were also showing up in the Deep South to support the new black-voter registration drives. Their presence was a profound criticism of the way Mississippi was run, and they were deeply resented in white-run towns with a history of violent politics. Buying the TV business in Tupelo, the company had also picked up cable rights in Laurel and two other Mississippi communities, which turned out to be in the middle of a guerilla war zone, as Brodsky and Aaron liked to tell it.[1]

"I looked like a civil rights worker, I talked like a civil rights worker," Brodsky said. "Laurel was the home of the Ku Klux Klan in Mississippi. It was boom, there goes a church. Boom, there's explosions. Boom, there's fires." Driving to Meridian, another town they hoped to wire, Roberts, Aaron, and Brodsky made the acquaintance of a real Jewish civil rights worker, young Michael Schwerner, later killed by Klan sympathizers and buried with two colleagues in a dam.

But the cable guys weren't down there to overthrow the local power structure; they wanted to have dinner with it, and do a deal. The social change they brought was cultural and subtly subversive, not political or openly revolutionary. They might not have been pleased that there were ugly forces working in this land; but they wouldn't likely have had this opportunity if Mississippi had been peaceful, quiet, and welcoming to more substantial investors. Tired of hanging out in small-town restaurants like Weidman's in Meridian, where they disliked the wheat salad and corned beef so much that Roberts made a habit of cleaning out the complimentary peanut butter jars (which the help learned to hide), the trio moved on to country establishments like the Queen of Hearts roadhouse—the kind of nominally private club you had to haunt if you wanted to drink with the local gentry. Visiting the Queen, Roberts asked a craps-table player about his Ivy League accent, learned he was the Meridian city controller, and found out all he needed to know about whom to approach for the city cable rights: one Ronald Goodling, owner of a trucking company, who cheerily told Ralph everything he owned was for sale, including his daughters, and who did, indeed, sell him his newly acquired cable franchise.[2]

Hiding from the Customers

Of course it couldn't be that simple; and it wasn't. The franchise had other local suitors; Meridian required a $125,000 bond as a condition, to make sure Comcast would make good on its outrageous promise to wire 90 percent of its applicants within a year. "Meridian City Council was so sure we would lose the bond, they allocated our money for the purchase of voting machines," Brodsky said. But Roberts's investors back at Wolf, Block in Philadelphia guaranteed the bond—and Aaron came up with a way they could keep their money by fixing things so the company wasn't swamped with applicants: "Let's hide."

Aaron moved the company's Meridian office out of sight atop the local Dreyfus family's Three Foot Building. (The Philadelphians knew enough Yiddish to get the joke: *drei foos* means three foot.)[3] The office and the trucks carried no sign; Aaron hired out-of-town installers and college kids, and warned them to keep quiet about their mission. Roberts even claimed the crews worked at night, so homeowners wouldn't see them. And that's how they wired the neighborhoods the company most wanted to serve: at the company's own pace, instead of the public's, and within the letter, if not the spirit, of the terms of their contract. They were proud of their deception.

There were court challenges, and an attempt to cancel the company's bond insurance. According to Aaron, Roberts hired the best-connected old flowing-haired Mississippi lawyer in town to parry all threats, genuine and ludicrous. When rain shorted their signal, reducing customers' pictures to staticky snowstorms, Aaron recalled sending out his truck crew with broomsticks, with instructions to "beat hell out of the cable." Told a local manager was a bigot, he fired him; otherwise he relied exclusively on local managers, determined to minimize any backlash against Northerners like himself. Aaron sent the cleanest-cut college students he could find door-to-door to pick up customers in targeted neighborhoods. In the end, Roberts's team outlasted all opponents, and the Meridian City Council upheld his franchise.

In their office on City Line Avenue, the strange pairing of tall, deep-voiced Brodsky and little, quick-moving Aaron provoked pop-culture comparisons: Laurel and Hardy; Abbott and Costello. The combination got noticed as the little company grew: "You remember the old comic strip, Mutt and Jeff? The big guy and the little guy? Aaron was like Jeff, you wanted to put your arm around him and shield him from the outside world. Brodsky was Mutt and then some, he'd scare you with his bulk," recalled Philadelphia investor Theodore Aronson. They bickered, threatened, and went to lunch together. Roberts drew both of them out, then decided for the group.[4] Compared to the highly individualistic businessmen who were often their main competitors in the struggles to win new territory, the Roberts team—given their diverse skills and sober checks and balances—had a real advantage.

Useful People

The circle of Ralph's friends and relations did a heroic job of raising money for his first TV venture. But the network wasn't big enough to sustain a growing company, especially if Ralph expected to keep control. Most cable operators turned to commercial finance companies—high-risk, high-interest lenders used to collecting their money the hard way from small-time clients. Many cable investors bought and sold systems so quickly they had little interest in permanent financing. Foreclosure was always an option for those who didn't want to go too deep. But even from the beginning, the Roberts group was building to last. Ralph still had friendly loan officer Jack McDowell over at PNB, but even he couldn't come up with all the long-term, low-rate working capital they craved. Cable was too new.

As always the group used personal connections, or made new ones. At one of his New York clubs, Ralph's brother Joe talked an executive at Home Life Insurance Company into meeting Ralph. Insurers in the mid-1960s were beginning to look for creative ways to deploy the surplus capital they collected in that era's lively securities markets.[5] As Brodsky remembered it, the executive showed little interest during the meeting. But he did read the prospectus they left him, full of promising reports on their Mississippi cash flow. That piqued his interest, and he called back to continue the conversation.[6] Home Life bought in; in time, so did industry giants Teachers Insurance and Annuity Association (TIAA), Mutual of New York, and Massachusetts Mutual. As other cable companies launched costly and desperate schemes, Brodsky and the Roberts brothers latched onto a respectable, patient, and unusually affordable source of funds.

Cable Goes to Town

As long as it was relegated to places like Pennsylvania coal country and Mississippi's Delta country, cable would remain a marginal business. But cable attracted visionaries who thought big almost from the start, and they were soon at work hatching big plans to bring cable to the nation's cities and metro areas where most TV watchers lived. TV was as free as radio; their job was to find ways to stick a meter on it, and get people to like it enough to put up with their monthly bills.

Irving Berlin Kahn was the trendsetter. He was the first cable executive to win national prominence, among the first to split a big-city franchise, one of the first to attract big corporate money; he set a trail many smaller operators followed. He was also the first to go to federal prison. Kahn was named after his celebrity uncle, singer–songwriter Irving Berlin, a Russian Jewish immigrant who penned "I'm Dreaming of a White Christmas" and popularized wholesome pre-electric pop songs in an active career that started before radio and stretched into the television era, from composer George Gershwin to TV showman Ed Sullivan. It was an irony of that so-called Golden Age of Television that popular entertainment programs, produced in Manhattan, were invisible to some of the island's first TV owners, who saw local stations through an electronic blizzard on their boxy floor sets. Interference from office and apartment towers was so great that for broadcast purposes many of these viewers might as well have been buried in Pennsylvania coal country. Many buildings could get clear reception only with complex apartment antenna systems.

Young Irving Kahn would change that. Kahn had the family show-business touch, but wanted to build an empire of his own. He had left a Hollywood executive's job at 20th Century Fox to promote such ventures as the prompting machine that helped President Kennedy (and blow-dried politicians ever since) look sincere by reading speeches straight into the camera; televised boxing bouts featuring heavyweight champ Floyd Patterson; even a new cable TV franchise in an old New Mexico mining town. He named his little conglomerate TelePrompTer Corporation, after his speech-reading machine. In 1965, Kahn won his biggest coup, a municipal contract to connect half of Manhattan through TelePrompTer. His franchise covered the poorer half of the island, with its miles of grimy, brick, six-story apartment blocks, but it was also the more populous section, and so more profitable. By the end of the 1970s, TelePrompTer was the nation's biggest cable company when Westinghouse bought it in the first wave of corporate cable mergers.

TelePromTer was also cable's biggest cash magnet during the go-go stock market of the 1960s. Speculative mutual funds and other big investors not only shoveled up the cash; they also buzzed happily about cable's prospects in the hero-worshipping financial press that develops in boom times. Kahn hopscotched around the country buying or displacing local cable monopolies, inspiring ambitious cable pioneers to embark on their own expansions and maybe "go public" too. Cable, like other emergent industries, badly needed a charismatic public face to excite investors, regale reporters, and, if need be, look good in front of a Congressional committee. "Irving Kahn was the industry's cheerleader, twirling his baton as he applied for franchises from Alabama to New York," Daniel Aaron recalled.[7] At a time when Roberts felt lucky to have the trust of a Philadelphia loan officer, Kahn was personally courting David Rockefeller, chairman of the mighty Chase Manhattan Bank, which became a major cable lender.

Kahn might have gone on to be the Bill Gates of the cable world. But the nation wasn't ready to endorse all of his tactics. In 1971, federal authorities charged him with offering city councilmen money for the Johnstown, Pennsylvania, cable monopoly. Convicted not just of bribery but also of lying about it, he was sent across the mountains to the federal penitentiary at Allenwood. His felony convictions didn't stop Kahn from reentering the cable business and developing New Jersey systems, which he later sold to the *New York Times* for $100 million. Aaron was among the cable executives who organized welcome-back events; in 1998, Kahn's peers named him posthumously to the Cable Hall of Fame in Denver. But after his 1971 conviction, securities regulators halted trading in TelePrompTer stock and demanded an audit, setting back cable's already slumping prospects.

TelePrompTer was acquired by Jack Kent Cooke, the Canadian high-school dropout who became an influential Washington businessman best known as the billionaire owner of the Washington Redskins and L.A. Lakers, setting a lucrative early example of the power of combining local cable and sports coverage. With his Washington contacts in everything from the Republican Party to the Central Intelligence Agency, Cooke was even better placed than Kahn to lead the industry. But Cooke wasn't a team player; for him, cable was just *a* business, not *the* business, and he sold—at a big profit—before his death in 1997.

Slicing the Pie

For Roberts's small but growing cable business, the stock market was still beyond reach. Brodsky turned his attention to partnerships. A mainstay of small-business expansion from early times, partnerships were also a favorite vehicle for the Dolan family's cable networks, which shared Manhattan with TelePrompTer, and which eventually held partnerships in systems all over the country. As Brodsky saw it, if Ralph could control his Mississippi cable systems with help from friendly hometown investors like Perelman and Wolf, why couldn't they bring in more investors to acquire other separate systems?

One of Roberts's most successful partnerships, for a time, was with the McLean family, owner of the nation's biggest afternoon paper, the sometimes chipper, usually bland Philadelphia *Bulletin*, which hoped TV would help it keep ahead of its archrival, Walter Annenberg's harsh morning *Inquirer*. The McLean–Roberts partnership helped Comcast acquire a smattering of suburban franchises near Philadelphia; next, the partners leaped over to the pleasant, sunny town of Santa Barbara, California, and to Sarasota, Florida, where Roberts had been outbid on an earlier attempt.

Indeed, Roberts had barely wired his Mississippi towns when the giant prize of Philadelphia came into view. By 1964, Roberts was already jockeying with Annenberg for rights to build Philadelphia's first cable TV system. Roberts hired David Berger, former city solicitor well known to city council's Democratic majority, and suggested that—instead of competing—the two companies split the city, each wiring half, as did TelePrompTer and Cablevision in New York. Instead of competition,

Solomonic division would ensure mini-monopolies for both. While city council weighed their plans, radio chains and out-of-town hopefuls rushed to file their own applications and hire their own politician–lobbyists. By 1966, Roberts and his rivals were willing to settle for carving their hometown like a pie. City officials responded by splitting the city into six franchises, enough to go around, while ensuring full employment for council lobbyists, at least until the contracts were signed.

But regulators in Washington intervened. The franchisers had hoped to pick up stations from New York and other cities, in order to have something extra to offer broadcast customers. The Federal Communications Commission, prompted by broadcasters, refused.

Then, blunt, charismatic Frank Rizzo succeeded James H. J. Tate as Philadelphia's mayor, leaving cable prospects even more remote. Cable had little priority on Rizzo's busy agenda, and Suzanne Roberts's old vision of public television as an instrument of civic uplift, which had flickered under Tate as the heir to the reform Democrats she had served in the fifties, went out. Of all the cable competitors that had hoped to wire the city, only Times Mirror Corporation managed to wire a few thousand homes, in South Philadelphia. Times Mirror struggled on until the FCC ban was lifted, so it could wire New York stations to its modest customer base. The rest of the city would have to wait.

The *Bulletin* alliance had served Roberts's early cable ventures, but a partnership is only as strong as the partners' commitment. McLean, like other large investors for whom cable was just a sideline, wasn't willing to stay so diversified in the business slump that followed the stock market decline of the late 1960s. Selling everything but his flagship Philadelphia newspaper in 1969, he forced Roberts to abandon Sarasota—the only time Roberts felt forced to sell a system, he later said. Roberts's other systems were growing, and the group was gaining confidence with its increased experience. Once again, as with his foray into golf clubs, Roberts's partner had let him down.

Ralph in Doubt?

It's part of the Comcast legend that Ralph Roberts never wavered in his faith in cable TV. But that's not exactly how his contemporary Harold Fitzgerald "Gerry" Lenfest remembered things at the end of the 1960s: The stock market was slipping; smart money was getting out of cable TV; and Lenfest was a busy young attorney at New York's Davis, Polk &

Wardwell, which counted powerful publisher and heir Walter Annenberg, as well as the upstart investor Roberts, among its Philadelphia clients. Indeed, Lenfest said that, as late as the end of the 1960s, when Roberts was still running a small conglomerate in which cable was one of several businesses, the Comcast founder was still far from certain about cable's future.[8]

"Ralph Roberts called me up around 1969 and said, 'Let's have lunch,'" said Lenfest in his modern Center City high-rise rooms, just across Rittenhouse Square from Ralph's hotel suite. "He said 'I'd like to put you in charge of our cable.' He believed Muzak would be a good business. But he said, 'I have serious doubts as to whether cable will ever be successful.'"[9]

But Annenberg was also interested in Lenfest, and the ambitious lawyer considered the older, wealthier, and better-established man's company a more promising prospect than what Roberts had to offer. Leaving Davis Polk for the Philadelphia publisher's headquarters, Lenfest was put in charge of an unlikely menu of businesses that included *Seventeen* magazine and the nascent cable unit; he rapidly got to know the leaders of the cable business, including New York's Charles Dolan, who in 1971 sent Lenfest a note on stationery from the giant Queen Elizabeth II ocean liner, outlining plans for the single invention that would make a lucrative national cable business possible: "a coded movie channel," which broadcasters and other freeloaders couldn't get, but cable viewers could, for a premium fee. It would be called Home Box Office—HBO.[10]

But HBO's early R-rated menu of sports, sex, and violence wasn't yet ready for prime time, and there remained the problem of staying in business at least until it was. Lacking Annenberg's (or McLean's) deep pockets, Roberts went back to work searching for ways to set his fragile empire on a firmer basis, without having to depend so much on the goodwill of others. Maybe he could, after all, sell shares.

Band of Brothers

Even as he moved toward taking his company public, Ralph still wasn't prepared to drop everything else and concentrate on cable TV. There was, for example, his big brother to keep in mind. The two had shared almost a symbiotic career: Joseph Roberts had risen through the marketing ranks at Revlon in the 1950s, pushing cosmetics, while his brother had expanded into men's toiletries, with Joe giving Ralph advice at every step of his entrepreneurial career. By 1965, Joe had switched bosses; he was running Muzak, the same comp·ıny Ralph had promoted as an ad man in the early 1950s, for its new owner, movie producer Jack Wrather, whose TV offerings included that heroic lawman, the Lone Ranger, and that heroic collie dog, Lassie.

Muzak's biggest customer was Storecast Corporation, which ran sound-systems at grocery chains. National Stores Corporation, the owner of Storecast, put it on the market; Wrather didn't want to buy, so Joe pitched the deal to Ralph, who bought with the understanding that Joe would quit his job and run it for Ralph from Joe's New York office on Park Avenue. And, the next year, Joe did. To Storecast, Joe added a string of Muzak franchises. Together, the business was International Equity's largest, providing more income than cable TV well into the 1970s. The combination placed the Roberts brothers squarely amid the forces working to promote mass-market production, consumption, and entertainment among the rising generation born after World War II. Muzak was originally designed to persuade laboring Americans to work harder in factories and offices; Storecast urged American consumers to buy more in chain stores; and Comcast urged them all to watch more television.

But the brothers' partnership was soon interrupted. Like their father, Joe died before his time. On a business trip to Spain, he developed what he initially thought was an ulcer. Joseph Wahl Roberts died of cancer in 1972.

CAPITAL

To Sell and to Hold

South of Chestnut and east of Broad, to the remains of the old Philadelphia financial district where J. P. Morgan once called weekly on his boss
A. J. Drexel,[1] tramped Ralph Roberts with his elegant overcoat, his hulk-
ing accountant, his squabbling advisers, and a couple of fat briefcases.

They were a little late. By that spring of 1969, the inflated stock market that had happily floated so many questionable attempts to fund "the
next IBM" was over; stocks would stay flat or worse until the 1980s,
when Ronald Reagan's election brought the cycle around and speculation became bankable again. But Ralph Roberts was determined to
draw new capital from the investing public, whatever the times. He
wanted to sell, not just stock, but also the bonds that only a stock-issu-
ing public company—certified, audited, government- and stock-market-
approved—could offer, at respectable rates of interest. They talked to the
Drexel firm, which in 1969 absorbed Gerstley, Sunstein, the Philadelpia
Jewish brokerage that had belonged to Suzanne's father and stepfather;[2]
to Janney Montgomery Scott; and to Boenning & Scattergood. And they
settled on little Supplee Moseley, whose bankers began a search for investors among the conservative inheritors of Philadelphia's nineteenth-
century fortunes in railroads, steel, and specialty manufacturing—not
always the folks most eager to try something new.

The bankers agreed that "International Equity" was too grandiose a name for the little collection of companies. Ralph settled on "Comcast," which both joined communications with broadcasting, and quietly balanced the Roberts brothers' interests—Ralph's cable and Joe's Storecast.

Giving It Away

Before a single share was sold, Ralph was giving the stuff away. Acquiring Westmoreland Cable Company in western Pennsylvania's richest county later that year, he pledged $450,000 in stock; those Pittsburghers who held the shares through the 1970s made tidy sums. But Supplee Moseley made little progress lining up cash buyers. Finally, Ralph took the deal back and brought it to Butcher & Sherrerd at the corner of Broad and Walnut. Young Butcher partner Joseph Castle, already a veteran of Jack McDowell's Philadelphia National Bank and later a Comcast director, handled the deal. The Butcher firm had taken Milton Shapp's Jerrold Electronics public and sold the firm to investor Moses Shapiro when Shapp decided to leave cable for politics; Butcher had a designated media finance group and cultivated the cable pioneers as clients.

The offering was set for June 1972. "It was a terrible time to go public," Brodsky recalled. Comcast listed on an off-brand regional stock exchange, and Butcher cobbled together a syndicate of brokers to retail the stock to their own clients.[3] The offering was supposed to go at $10 a share. But cable was weak. As *Barron's* inconveniently pointed out on the eve of the sale, cable financing from other sources had dried up; the areas most in need of the service already had it, and broadcasters were lobbying hard to keep cable at the margins of competition. So Butcher cut the share price to $7 and managed to sell the shares. And from there, as Brodsky recalled, "it fell like a rock."

The worst was the next three years. Comcast fell as low as 73 cents a share; or less than a dime in current terms, adjusted for subsequent stock splits; or a quarter, adjusted for inflation. Of course, "from 20,000 feet," as Brodsky liked to put it, $1 invested in the company at the bottom would be worth more than $100 twenty years later. But they didn't feel so smart at the time. After work, before unwinding a bit at Williamson's in Bala Cynwyd or some other nearby establishment, Brodsky recalled, "You had a decision: Do I buy a beer, or a share of Comcast?"

Even in the face of yawning indifference, Ralph wasn't willing to sweeten the deal by giving up control. He loved to consult; he was generally affable; he could be steely; but he must be in charge. Ralph insisted—and his allies on the board agreed—that he hold a majority of the voting rights, and veto-proof control. That was one thing, in that year of concessions, that Roberts refused to negotiate away, and retained as long as he was boss.

The Way We Live Now

By 1971, as Ralph moved to put his corporation in the firmament of corporate America, the world where he and Suzanne had met and begun their life together had largely dissolved. Philadelphia was a tense place, where conservative Mayor Frank Rizzo signaled reaction against the problem-solving do-gooders of the previous political generation, and called for nightsticks and patronage, not anything like public television, as a cure for social problems.

The Comcast partners had all attended city high schools—Brodsky at Overbrook, Aaron at West Philly, and Roberts up at Germantown—and thrived amid the social and academic challenges of those diverse neighborhood institutions, as well as through the more specifically Jewish activities they also attended. But now community life was shifting to the margins of the city and certain pleasant suburbs. Their children would live with expanded horizons, and aspirations all their own.

There were second-generation echoes of the immigrant past. For the Roberts sons, there were summers at Kamp Kewanee in the Poconos, where Jewish boys from what were by now comfortable Philadelphia families learned team and racquet sports and crafts in a structured program their fathers would have recognized. Big Brother Robbie (Ralph Jr.) went first; later, Brian and Doug tagged along. The Roberts home in Wyncote in those days was a showplace, full of antiques, recalled camp counselor Noel Stansell, who, invited back to meet their parents, pronounced both Ralph and Sue "incredibly artistic."[4]

By then, with her daughters already in college, Suzanne found herself casting about for a way to grab her family's attention and give them one more unifying experience that would help set their identity, just as her father and uncle had sought to impress on the young Fleishers their Old World ideals of duty and service long ago; and she found it, in the form of an improbable six-week trip to the nation's largest Indian reservation.

The Adopt-a-Nation Vacation

In vintage Fleisher style, Suzanne publicized the 1971 trip—she even issued a news release that referred to her husband as a "management consultant and president of Comcast Corp., a communications company"—and it's hard not to credit her hint that some of the social agencies she approached in Indian country were bemused by, even suspicious of, these wealthy tourists and their motives.[5]

But the Roberts family found its way, Mom reported. Everyone, it seemed, was true to type: Ralph Sr., entrepreneur, "helped start a Navajo language radio station" and invited women weavers to bypass their middlemen and sell directly to Wanamaker's back home. Cathy helped "two opposing teen groups" apply for a government grant. Lisa volunteered at a Catholic program for crippled children. Robbie did magic tricks at Indian health centers. These older three would soon enter the helping professions their mother clearly favored. Even the youngsters who would enter more combative lines of work, Doug the future lawyer and Brian his father's shadow, "helped Navajo grandmothers herd sheep" and, echoing just a little their father's youthful commercial urges, sold hot dogs at Indian rodeos, Suzanne reported. (Among the tribes, Suzanne and Ralph also went on a shopping spree, bringing home to Elkins Park bundles of Southwestern weaving and colorful handicrafts to complicate their previous Early American decor.)

Suzanne also found her next vocation: Volunteering at the Navajo hospital psychiatric ward, she claimed success at reaching withdrawn children "using techniques of rhythm and dancing." She wrote about her experience, convinced St. Christopher's Hospital in Philadelphia and a string of Scandinavian health agencies to let her demonstrate, filmed her techniques, enrolled for her master's in therapy at Antioch College, and volunteered as a therapist at a string of Philadelphia hospitals.

Meanwhile, Ralph, the advertised "consultant" who ran a "communications company" on the side, said much later that by this time he considered the cable business had become "dull as dishwater."[6] But that would soon change, thanks to the placement of TV antennas in places even Milton Shapp hadn't been able to reach—in orbit high above Earth.

Gentleman, Monopolist, Bear

The Rainmaker looked comfortable at his table in a paneled dining room not far from the Main Line of the former Pennsylvania Railroad. It was a fine post from which to render judgments on the gathering and dissipation of Philadelphia family fortunes and individual reputations. He'd seen so many grow and fade. In the 1960s, he numbered part of the crowd of Pennsylvania investors who traded in cable television franchises—not to keep them the way Roberts did—but for short-term gains, much as his predecessors and Roberts's in-laws might have swapped the lesser railroad bonds in an earlier day. As a group, he recalled, these Philadelphians were impressed by upstart Milton Shapp's success at making a fortune from selling cable equipment, then using the money to establish himself in politics.

But the Rainmaker's companions in those days were men with long memories, to whom the Crash of 1929 merely confirmed the lessons their fathers had recounted from the Panics of 1871 and 1893: the virtues of a sell discipline, for example, and not falling in love with uncorroborated stories even if convincingly told, and always, always keeping alert for an opportunity to sell the asset, take the money, and run, not walk, to the nearest exit. Even as they bought and sold cable partnerships in the boom years of the 1960s, most of them didn't think much of the long-term prospects for cable operators, whose main assets seemed to be quick turnover and tax breaks. They didn't have Ralph Roberts's growing faith in a cable future beyond the quick buck.

So how, you asked him, did Ralph and his team manage to go beyond mere trading and cobble together something as dominating as Comcast?

The Rainmaker praised Roberts's ability to remain "a gentlemanly, dignified" sort in social settings, yet "a bear—to quote [Ralph's] own words" at business. He admired Comcast financial wizard Julian Brodsky's eye for making cunning investments—he might have noted Philadelphia's moonshot Internet Capital Group, a late 1990s example—and for taking profits before they blew away.

Goaded, the Rainmaker went further: "They have played a very close-to-the-vest policy all the time, and kept themselves ready to take advantage of the silliness practiced by so many other managements, such as AT&T.

"They have minded their own business exceptionally well, and cultivated lots of good PR from the local flacks.

"They are very smart, to boot.

"Fundamentally, though, they are monopolists. Their customer service is — was — notoriously bad. They have been raising prices inexorably over the years, and under present regulations I see no end to it."

And then he dismissed Comcast, and you.

6
SATELLITE

Dirty Movies

It took help from outer space to bring cable TV from America's isolated margins to its lucrative mainstream. To grow beyond the boondocks, cable needed something broadcast TV couldn't or didn't dare to offer. Home Box Office provided that something, with help from a series of satellites orbiting above the United States. Once HBO hit its stride, as cable magnate John Malone put it, "I [could] watch a dirty movie anytime of the day or night."[1]

In the early 1970s, long before *Sex in the City* or *The Sopranos* kept too many Americans on the couch watching actors break the Ten Commandments, HBO was a little offshoot of a debt-ridden New York cable company, trying to convince small-town Americans to pay extra for little boxes they could wire onto their TV sets, which would allow them to watch polka dancers, B-list sports, and retread movies for a surcharge on their monthly cable TV bills. It wasn't exactly glamorous.

HBO was the brainchild of Charles F. Dolan, a maker of films for business, who had convinced Time Incorporated and other big investors to finance his bid to wire the half of Manhattan that Irving Kahn didn't get through Dolan's company, Sterling Manhattan; he didn't have much in the way of profits to show his investors. HBO would give his customers something to watch besides broadcast TV; its first cable programming included pro sports that would otherwise have been blacked out by broadcasters, plus second-run movies. At first Dolan gave HBO away, but he quickly found customers were willing to pay $6 a month to get more than their neighbors. That was an impressive sum, considering that most people still thought of TV as free, like radio, or air. It was expensive, but the concept intrigued Time, and a young Time lawyer, Gerald Levin, who elbowed Dolan out of the way and went on to make Time Warner one of the nation's dominant media companies.

Dolan went on to greater success as head of Cablevision Corporation, which became New York's dominant cable and sports TV company, while helping to finance many of the biggest cable systems across the country. There was wide and deep interest in pay TV's success: Cablevision's early backers included Milton Shapp's old Jerrold Electronics, which hoped to sell plenty of cable boxes; Milton Friedman, the supply-side economist, who lent respectability to President Reagan's tax cuts for the rich; and Hugh Hefner, the porn merchant, who, like plenty of the cable operators carrying his Playboy Channel, never let his company's chronic lack of reported profits prevent him from living like a very wealthy man.

Look, Up in the Sky

Time Incorporated wasn't content to wait until programs could be wired around the country. At first, HBO used microwaves to bounce from system to system. That was a step up from traditional broadcast antennas, whose use was intermittently restricted by the Federal Communications Commission, at the behest of the broadcast TV industry and its allies (usually Republicans) in Congress. John Walson's Service Electric Company, which had expanded over twenty-five years from little Mahanoy City to slightly larger towns like flood-prone Wilkes-Barre in the next county north, had been HBO's first out-of-town user; it carried a New York Rangers game and actor Paul Newman's portrayal of a strikebreaking logger in the movie *Sometimes a Great Notion* to a few hundred homes late in 1972.

Two years later, Jerrold's main competitor, Atlanta cable-equipment maker Sidney Topol of Scientific Atlanta Corporation, brought HBO together with satellite maker RCA and a group of cable operators to talk about taking the service national. Among the interested parties was young R. E. "Ted" Turner, who ran an independent broadcast station that showed a lot of old movies he was itching to take to a national audience.[2] RCA had recently won federal approval to launch a new communications satellite for bouncing signals between cities from space; Topol proposed using it—and his company's satellite signal stations—to beam HBO programs to local cable systems, giving them something broadcast networks didn't have, with which they could woo viewers even in places where home TV antennas worked fine. (Theater owners fought back for a time, sponsoring a "Save Free TV!" ad campaign that helped create a climate in which it was hard for HBO to get its hands on the most popular as well as the most recent films.)

County Fairs and Heavyweights

While RCA got ready to launch its satellite from the government's Florida space center, HBO struggled with its attempts to provide original programming on a shoestring. It bounced from New York hockey games to Pennsylvania county fairs to warmed-over movie nights until 1975, when it managed a last-minute hookup linking the heavily promoted "Thrilla in Manila" championship boxing match between Muhammad Ali and his Philadelphia nemesis, Joe Frazier, to American cable systems.

The signal was beamed back through Topol's connections through a more primitive satellite, Western Electric Company's Westar I, which looked like a pregnant oil drum with a Little League pitching machine bolted to the rim. By the time Ali finished outdancing and drubbing Frazier, HBO had delivered a headline event from the other side of the world, something you couldn't come close to seeing on broadcast TV.[3]

Now cable operators all over the country were interested. The next year HBO switched to a more sophisticated satellite, RCA's new $40 million Satcom I. Satcom looked like a flying kitchen range, with two waffle-shaped solar panels sticking out the sides and a dish antenna stuck on the back. But its size and reliability helped convince Madison Square Garden in New York to let HBO beam live sports and concerts to markets across the country on a regular basis; in Hollywood, Universal Pictures signed a deal allowing the use of its vast film library. By the end of the next year, HBO had more than 1.6 million subscribers—up from 8,000 three years earlier—all paying $8 a month, plus $30 up front to buy the little box that unscrambled signals.

Comcast declined the first invitations to carry HBO. Roberts wasn't crazy about R-rated movies; still, like everyone else in cable, he loved the idea of having a service he could sell that broadcasters couldn't offer. But, with wealthy companies like Time and Westinghouse throwing millions around, Roberts waited to see if urban customers were really willing to pay more for the extra service. Finally, in 1977, he began offering HBO to his customers in suburban Pittsburgh—and looking for ways to start marketing this new programming channel, and others like it, to urban customers he had bypassed until now.[4]

Gold Rush

HBO made cable attractive to mainstream investors once more. This time it was some of the nation's biggest corporations that came calling, bearing large piles of cash.

"Cablemania is here again, promising that this time it will avoid the debacles of the Sixties and early Seventies, when brief cable booms were followed by big cable busts," Allen Sloan wrote in *Forbes*.[5] "Instead of your old-time cable touter—a promoter leveraged to the eyeballs, holding his company's stock together with baling wire and accounting tricks—cable is attracting corporate America's version of respectable old money."

In the late 1970s, American Express Company paid $175 million for half of Warner Brothers' cable network. General Electric offered nearly half a billion for a stake in the Cox cable network; and Time paid $140 million for a controlling stake in another system. Banks, insurers, and equipment makers kept busy writing new cable business. The New York Times Company, Knight-Ridder, and other newspaper chains mulled their own cable investments after the owner of the *Los Angeles Times* invested over $100 million in a string of local cable monopolies. There was certainly room for someone to grow. Before HBO, cable systems in the biggest cities—in New York and Los Angeles—had lost money. The next biggest—Chicago, Philadelphia, Houston, and Detroit—still weren't wired for cable; it didn't seem profitable, with basic cable rates, still regulated by city councils, leaving little room for hope of profit.

HBO changed all that. "A cable operator scraping by, selling basic cable service at, say, $6.95 a month, suddenly had HBO to sell at, say, $8.50. He kept $4.25 and sent the other $4.25 to Time," which had rented the RCA satellite for $1.2 million a year. "The $4.25 was almost all pure pretax profit for the operator; the wires were already in place." Sloan figured Time was grossing $150 million a year retailing the service to customers and other cable systems.

That sudden profitability created huge temptations. Sloan calculated the Pittsburgh cable system was worth around $36 million, based on estimates by the companies competing for the franchise. Yet Comcast and TelePrompTer sold one-eighth of the venture to a group of local partners for less than $300,000. Time made a rival offer to sell 20 percent to another group of influential locals for just $500,000. "It's all perfectly legal. The money virtually being given away to local investors in return for political backing dwarfs the $15,000 former TelePrompTer Chairman Irving Kahn was convicted of bribing city officials with in Johnstown, Pennsylvania. Kahn was sentenced to five years in jail for bribery and perjury. There is," Sloan complained. "an element of we-will-win-at-any-cost."

At the time, *Forbes* listed Comcast as the smallest of six publicly traded cable companies—all of which, Sloan's sources insisted, would surely be bought by larger companies during the onrushing boom to wire the nation's seventy-five million television-owning homes.

Rent-a-Citizen

The cable pioneers had always resented having to deal with local politicians. "Prior to deregulation, rates were regulated by city councils," Comcast's Brodsky recalled, with some disgust. "You actually needed permission to raise your rates."[6]

Cable people called it "rent-a-citizen," the game in which rival cable networks raced to hire influential lobbyists or invite influential citizens to buy a share of new cable ventures at sweetheart prices in order to win a coveted franchise. As far back as his early visits to Mississippi, Roberts had understood the importance of making friends among the local power structure. Now Comcast assembled a special team with all the skills required to barnstorm likely targets during the wars that broke out as each community in America, over a span of twenty years, gave out its initial cable franchises. Comcast's team tried to cover all the bases: "We had Barbara Lukens, a Wellesley grad from a Quaker family in town; Ed McGuire, an Irish politician; and Abe Patlov, who knew the cable business." Finesse, clout, and depth. All their talents were needed, and repeated, as Comcast built its network.[7]

Brodsky said he never knew the level of intrigue to fall to the level of violent threats at the hands of local factions, as some of his rivals have claimed. Still, "I had to put in so many safeguards to make sure no cowboys did anything with bribes. On any payment of $5,000 or more we had to get two signatures by officers involved in the deal."

Why so careful? "A number of guys in the industry went to jail." Irving Berlin Kahn's imprisonment for bribing officials in Johnstown,

Pennsylvania, to renew a franchise struck Brodsky as especially sad: "It was a tragedy. He was our charismatic leader. He looked like a Hollywood mogul."

A generation later, with Kahn gone to his final reward or punishment, Brodsky declined to pass judgment on what he went to prison for. But he insisted Comcast always stayed clean, walking away from deals that crossed the boundary into criminal territory: "Our proudest moment was that we went 0 for 22 in Massachusetts."

Spreading Out

Half for Me?

Ralph Roberts didn't throw money around. He was an opportunist who couldn't see paying a lot when he might wait and pay a little. And that was just fine with his rivals, during the long contest to lock up small-town cable monopolies.

"I had a motto: 'If you don't buy it you won't own it.' And that was not Ralph's way," recalled Harold Fitzgerald "Gerry" Lenfest, who competed with Roberts for cable deals around Philadelphia. "I remember the day we bought Norristown Cablevision of Pennsylvania. Ralph called me and said, 'How the hell could you pay all that money?' Half an hour later he called back and said, 'Can I buy half?'"[1]

Lenfest won most of his battles with Roberts, except his last one, when Comcast won the war and forced Lenfest to sell what he had spent thirty years building; he wasn't the only one whose long, friendly wrestling match with Comcast ended abruptly, with him outside the ring. Though it's hard to say Lenfest lost: He walked away a billionaire.

But that came much later. By 1980, seven years after Comcast had gone public and seventeen years since it wired its first TV, Comcast was still a marginal player in the national cable business, dwarfed by big corporations with more money and better connections in Washington, Hollywood, and the city halls and county courthouses where the biggest cable deals were getting made. The new TV satellites had attracted another wave of fortune-hunting investors, including big corporations that offered premiums and sweeteners to land the biggest deals.

> Pictures so clear, they're better than real!
> Public access studios, free to all!
> Home games for your favorite pro team!
> Grown-up movies—uncut!
> Seventy-two channels!

What would those crazy cable guys think of next? And how could a family business hope to compete?

Comcast trailed; its stock price was still underwater, worth less than investors had paid for the first shares. They would have done better with their money in the bank—or a mattress. As late as 1977, cable still accounted for less than half the company's sales; canned music and canned shopping announcements were still its unglamorous main sources of revenue, and Ralph Roberts wasn't ready to let them go, or to make cable commitments he thought were too generous.[2]

The Also-Rans

When Comcast in the 1970s faced level-playing-field competition for a really first-class cable market, Roberts and his lieutenants had a tough time competing.

Like an heiress weighing marriage proposals, the wealthy New York City suburbs of Fairfield County, Connecticut, spent more than a year choosing which of ten suitors would wire more than 100,000 homes in wealthy Stamford, Greenwich, and neighboring towns, just across the state line from Ralph's boyhood home in New Rochelle. Home to General Electric Corporation and other corporate refugees from the surly, nearly bankrupt Manhattan of the 1970s, Fairfield promised to choose the classiest connector for the right reasons. "The Fairfield franchise will be at least the best system in the state, and quite possibly the best in the country," state consumer counsel Barry Zitsner bragged to the *New York Times*. There would be no after-hours clubbing where dealmakers like Ralph could collect inside information over a craps game, no political carve-up where he and his rivals could hire connected lobbyists and lawyers to split the territory into lucrative, parallel, noncompetitive colonies. Instead, Connecticut's public utilities commission and consumer advocate said they would choose a provider in the best sensible, boring, open, good-government style.

Connecticut publicly rated and berated its suitors. The leading contenders included affiliates of the Dolan family's Cablevision; the cable units of the giant United Artists–Columbia studio in Hollywood; the Scripps–Howard newspaper chain; and two efforts backed by partners of Storer Communications, a Miami operator that had forced Roberts out of a lucrative Florida franchise, and a Canadian cable chain. At the bottom of the list were two small companies, which, though based in Connecticut, were utterly unable to exploit the process through their local connections, along with Comcast, hampered by "lack of money or the ability to match the offerings" of the front-runners, the regulators smugly reported. A Cablevision group got the nod.[3]

Freeing the Fortune 500

That fall, Ronald Reagan was elected president with a mandate to reverse the drift of American government over the past half-century. Reagan's program—less regulation, more taxes—trumped Jimmy Carter's moralistic appeal. The telecom industry looked to Reagan to extend what had begun under Carter: wiping back, not just the crust of regulation that had limited corporate power since Franklin Roosevelt's New Deal, but also the limits on business combination left over from Theodore Roosevelt's Progressive era.

Though Ralph Roberts and many other leading cable executives had started life as liberal Democrats, as they grew successful, the Reaganite ideal of less regulation—and the Reaganite reality of more corporate involvement in regulation—gained appeal. Over the next twenty years this became the dominant Washington view. In lobbying appeals and congressional committees, big media companies used the language of freedom to block limits on their power, curb their competitors, and arrange markets so they could charge higher fees than a truly open market would allow. Some of the most powerful figures in the communications business would grow wealthy through mergers that would have made old Teddy Roosevelt howl, with the compliant approval of the very agencies—like the FCC—that Franklin Roosevelt's brainy young men had set up in hopes of limiting corporate power.

But before they could build the next generation of business empires, the Reaganites faced a crushing problem: the high price of money. Basic interest rates in those days topped 20 percent, and couldn't be brought down quickly, because inflation fighters at the Federal Reserve didn't seem to answer to Reagan any more than they'd answered his desperate predecessor's begging to cut rates. If Wall Street couldn't get cheap bank loans, it would make respectable the next best thing: Debt—even high-risk, high-priced debt—would oil the engines of financial speculation and economic growth for small but aggressive companies like Comcast. For many companies this was an annoyance; like credit card users, they borrowed as if they cared only what the monthly payment was, not how their overall debt was rising. But for Comcast, debt in its bewildering creative new forms was an opportunity; applying the same careful tactics the company used in acquisitions, its debt financing would give Comcast an edge.

Other People's Money

In December 1981, Comcast was back on Philadelphia's Walnut Street, looking for cash. The firms that had backed Roberts at his riskiest, bound partly by old ties to the Fleisher family and the Philadelphia Jewish community, were now joined by aggressive mainstream firms looking to profit from the company's growth: Butcher & Singer, the investment firm that had taken Comcast public; Provident National Bank, which had grown bolder than old Philly National; and the law firm of Ballard, Spahr, Andrews & Ingersoll, now, like Wolf, Block, a Comcast counselor. Together, they helped Roberts convince Lower Merion Township—Pennsylvania's richest community and the company's home base—to lend local government sanction and tax breaks to a $3.5 million bond issue. It was quite an achievement: Comcast could wire the township with borrowed, tax-exempt funds, and pay it back from citizens' cable bills.[4]

Lower Merion, with its wealthy Main Line villages, was just one of many towns that ultimately allowed the company to use its financial good name to secure cheaper funds. And those bonds were just one of the weapons in Julian Brodsky's wonderfully diverse, flexible arsenal. There were Delaware investment holding companies, which converted income from distant franchises into "fees," avoiding local income taxes in more than a dozen states. And there were convertible securities in which bondholders were repaid, not with cash, but with Comcast stock, providing the company with a powerful incentive to get the price up before angry investors rebelled.

Comcast hired Butcher, and eventually bigger firms, to help retail one of Comcast's most characteristic products in those days: "safe harbor leases," which let Comcast trade its tax losses for cheap loans. *Business Week* published a disdainful "how-to" guide, in which Comcast served as Exhibit A.[5]

"A group of investment partners supply the investment kitty" to buy local cable equipment, which is then leased to operators like Comcast, *Business Week* explained. "The tax benefits—which attract most limited partners far more than the admittedly uncertain prospect of profit—include a five-year tax write-off" of the equipment, plus a federal tax credit, which increased as the enterprise bought more equipment.

As in *The Producers*, the Broadway show in which a ludicrous Nazi musical is calculated to make money only if no one goes, Comcast's tax shelters made money only as long as each cable system reported a loss. If it accidentally showed a profit, the Internal Revenue System could claim a cut. The magazine labeled the shelters "troublesome," and "dangerous."

But they sure proved popular. From local firms like Butcher, Comcast moved its cash-raising efforts to Wall Street and eventually established itself in Europe as well. Not to junk bond central—Drexel, Burnham, Lambert, whose clients under Michael Milken included Cablevision, and John Malone's TCI, in addition to other A-list cable monopolists—but to firms like Shearson/American Express. A somewhat more reputable outfit, Shearson had as current and past bosses, respectively, Sandy Weill, future architect of controversial Citigroup, the world's biggest bank, and Arthur Levitt, who, as Securities and Exchange Commission chairman, would approve the final destruction of the Depression-era laws that had carefully separated banking from investments. Levitt and Weill had been partners at the fertile and aggressive 1960s firm of Cogan, Berlind, Weill & Levitt, known to its detractors as "Corned Beef With Lettuce" in a snide cut at its partners' shared Judaism. In reports that were avidly read by a generation of bright young men on Wall Street, Weill's and Levitt's firm had projected both the junk-bond craze of the 1980s and the multi-billion-dollar financial-merger mania of the 1990s.

Shearson raised $25 million for Comcast by selling tax-protected limited partnerships in 1982, even though the prospectuses typically didn't even tell investors where the systems they were investing in were located. The next year, Comcast reached further and convinced Shearson and the venerable British brokerage Morgan Grenfell Company to join Comcast's new international-finance arm in backing Comcast franchises in the United Kingdom. Comcast would attempt a global reach.

MUSCLES

Crown Prince

Days after Ralph Roberts named his son Brian chief executive in 2002, Time Warner head Richard T. Parsons introduced Brian to a hall full of media moguls as the "crown prince" of Comcast, adding that Brian's infant foreskin had been cut off his penis with cable splicers during his Jewish ritual circumcision.

What could Brian say? "Geez, Dick, nice."[1]

Few human acts are more basic than a father trying to pass on the powers and properties he's amassed over a lifetime to his son; and few events are less common, today, in big public companies, where the typical CEO doesn't keep his job much longer than the typical starting NFL quarterback. That doesn't stop some from trying, especially when they have founded the company and still own a big chunk, at least of its voting stock. Rupert Murdoch provoked blistering criticism from his British shareholders when he put his sons in key positions at News Corporation. Latter-generation Fords and du Ponts, New York Times Rosenthals and Bank One McCoys, have faced more than their share of second-guessing in times of crisis over the years: Were they really the best men available for the job?

"People are going to assume the worst of you, that you were born with a silver spoon," Brian Roberts once said, in regard to American dy-

nasties. "If that's going to ruin the fun, and kill the joy and focus to suc-
ceed . . . then don't do it. You have to become comfortable in your skin."[2]

Brian Roberts has been spared most of that, partly because he really
did grow up as the committed son of a respected and feared father, amid
the strange fraternity of cheerful cutthroats that owned and ran the pay
TV business. But Brian's inheritance, like Jacob's in the Bible, was not
assured from birth. He was the fourth child, the second son, in the pre-
cocious Elkins Park household of Ralph and Suzanne, soaking up the
same rich atmosphere of culture and psychology, commerce and sports,
as his brothers and sisters. According to his father and his top lieutenants, Brian was simply the only one of the five children who took
much interest in what dad did every day, and often late into the night, to
support them in private-school, long-vacation comfort.

Ralph was still the Suspender King when Brian was born in 1954.
His second son was in kindergarten when Ralph bought his first cable
system, and Ralph made a point of not taking his work to the family dinner table. Brian alone seems to have sought his father's world. As Ralph
has told it, Brian relentlessly pushed for a place in the company even as
his father tried to warn him off. Ralph said he relented only when Brian
opened the psychological throttle, pressing to be with his father while
there was still time (as Roberts men like his uncle and grandfather had
died young); complaining even that he, Brian, felt rejected by his own
dad. Yet Julian Brodsky also said the matter was settled, more or less,
while Brian was still in grade school—by which time the kid was already
collating bills around the office for a quarter every Saturday morning,
and demanding more of both responsibility and pay.

As a student at elite suburban Germantown Academy—not to be
confused with the big-city public Germantown High School where dad
went—Brian did well enough to earn admission to Penn's Wharton
School, Ralph's alma mater. Starting college at 6' 2", but only 125
pounds, the future NBA owner bypassed team sports and went out for
squash, which he'd learned at summer camp. Brian "had a pretty stroke,
but an opponent could get in his head and provoke him to the edge of
tears," wrote Philadelphia sportswriter Michael Bamberger. Penn coach
Al Molloy called skinny Brian "Muscles," and benched him plenty, but
he also pushed him to lift weights and scrimmage against ex-Phillies
slugger Richie Ashburn; in time, Brian won the key set in a tough Harvard match. Molloy taught Brian "you have to be tough, physically and
mentally"; Brian told Bamberger, "Aside from my father, no one has in-
fluenced me more than Al Molloy."[3] Brian also took quickly to golf and
skiing, gentry sports that can be used to make points with business associates, and his squash prowess later won him three silver medals at the

Maccabiah games, the Jewish Olympics. He continues to play in nationally ranked amateur tournaments, along with his own teenaged son, Tucker. Brian has helped investor Warren Buffet beat Microsoft's Bill Gates on the tennis court, refused to let Buffet beat him on the golf course, and bested par on nine holes with Arnold Palmer. He has gone out of his way to challenge subordinates to foot-races on management retreats, and competitors to skiing competitions at industry gatherings—and beaten them.

Growing into It

As a teenager, Brian was as avid a business-watcher as his orphaned father had been a canny salesman. Brian haunted his father's office; he watched negotiations with an attentive silence, then quizzed his father relentlessly.[4] Not that it was an ascetic existence. In his memoirs, Comcast executive Dan Aaron accused Brian of using his office visits to swipe the *Playboy* magazines Aaron kept in his desk—where Aaron claimed he'd dumped them after confiscating them from his own teenagers. More important, Ralph's tutelage allowed for his son's mistakes. "If you don't criticize your children, they grow into it," Ralph said. Brian told the *Inquirer* that his father never told him, "That's a dumb idea." Instead he would say, "Have you thought about it this way?" In Roberts's family school of management, self-confidence came first.[5]

Already in the 1970s, Brian was sitting in on deal negotiations. "The thing he learned is that you can negotiate with a major lender. You don't have to fall on your nose first crack out of the bag," Ralph recalled thirty years later. Keep quiet, Brian was told, you might learn something.[6]

The Robertses have allowed reporters to claim Brian worked his way up from installer, lineman, door-to-door salesman, bookkeeper, and management trainee, so that no job among Comcast's sixty thousand would be strange to him. "Prior to his management responsibilities, Roberts worked part-time for five years in cable television sales, installation and construction," Ralph announced in a 1990 press release proclaiming Brian's elevation, at age thirty, to the company presidency. But there are signs Brian's blue-collar exposure was more an exotic life experience, like the family's earlier social-work vacation among the Navajos of New Mexico, than an old-fashioned trade apprenticeship.

During his last high school summer, Brian headed out to Colorado, where his big brother was in college, to tag along with a Comcast installation crew. It was an eye-opener: "Dad, you should see these guys," downing beers and pinching waitresses, Ralph said his son told him on one call home. However, in May 2003, at a Philadelphia cable equipment show, Brian—in the tradition of his father's comic self-deprecating tales—gave this version of his brief stint in the trenches: He was working with a veteran lineman when the cherry-picker bucket accidentally touched a live wire, electrifying the lineman. As the ambulance rushed the injured worker to the hospital, Brian "called his dad and told him about the accident, but said it was exciting to be an installer. The next day Roberts was reassigned as a door-to-door salesman," according to the account posted by the *CableFax* newsletter.[7]

Show the Boss How

The next year, Brian spent his pre-Wharton summer of 1976, not splicing cables or ringing doorbells, but as an intern in the air-conditioned Center City Philadelphia offices of stockbroker Drexel, Burnham, Lambert. The firm was the product of a recent mixed marriage between the faded Main Line Philadelphia firm originally known as Drexel & Company, then run by the formidably named Horatio Gates "Terry" Lloyd III, to the upstart, predominantly Jewish, Burnham & Company, which needed Drexel's white-shoe contacts to flex its new investment banking muscle. Drexel was where another bright Wharton student, soon-to-be junk-bond king Michael Milken, had spent his own internship a few years earlier;[8] he was now a rising star in the firm, building the career that would make him famous, exalt and wreck Drexel, Burnham, and send him to prison at the end of the 1980s.

Brian didn't walk in off the street to apply for a job. His internship was sponsored by Leon Sunstein Jr., whose father and step-grandfather had founded Gerstley, Sunstein & Company, where Brian had spent a few weeks helping his patron, who was his step-uncle (Leon Sr. had married Suzanne's mother, the widowed former Selma Gerstley Fleisher).[9] Comcast had tapped the rival crosstown Butcher firm to sell its first public stock offering three years earlier. But it was going to need to raise a lot more money to fulfill Ralph Roberts's expansive designs, and Drexel, Burnham had reason to hope it might eventually share the company's investment-banking business. So family and professional needs combined when Sunstein recommended Brian, his nephew and the Comcast boss's kid, as a Drexel intern to whoever could teach him what Wharton couldn't quite. (Drexel never did get Comcast's business. "We used Merrill and Morgan to sell high-grade bonds. Not Drexel" or its junk bonds, Roberts's financial wizard Julian Brodsky said years later. While "TCI and Turner were creatures of high-yield debt, I barely knew Mike Milken."[10])

At a Drexel staff meeting, young Roberts was offered as an intern to anyone who needed one. He was quickly chosen by a boyish money manager and fellow Penn grad, Theodore Aronson.

"We've Been Comcasted"

Today Aronson is senior partner in the several-billion-dollar firm of Aronson, Johnson, Ortiz, which counts Ivy League institutions among its clients. He is a professional skeptic and a quant: he picks companies based not on what, how, or to whom they sell but purely by the numbers. Asked how often he visits the CEOs of the companies he buys, he threw back, "Why should I talk to them? They lie!"[11]

But Aronson has been a Comcast cheerleader ever since that long-ago summer, based on his intern's two improbable accomplishments: Brian goaded Standard & Poor's into listing little Comcast for the first time in its Bible-like stock guide through what Aronson described as pure persistence. And Brian boldly, tenaciously, and successfully presumed to teach Drexel senior partner George Connell (future owner of Philadelphia's Rittenhouse Trust Company) a few points on putting, in Connell's Drexel office.

"I walk past. Brian is in the managing director's office. He has the putter. And he's just seventeen," Aronson marveled.

He is reminded of his old intern every time he walks into his Main Line living room: "I have Comcast high-definition TV, but I can't see the difference. I have Comcast video-on-demand; I can't work it. I have Comcast high-speed Internet. I am paying for all this. They set the price. We are Comcasted."

And, yet, Aronson the skeptic melts in the warmth of his personal embrace of Ralph and Brian, their long performance as the relative gentlemen in a pool of sharks: "I have followed his family's career with a brotherly affection. And every time I see or read about Brian Roberts on the first page of the *Wall Street Journal*, I cheer."

But seriously: What's to like about a family of price-gouging, wanna-be monopolists?

Just consider the competition, Aronson replied. "For years you'd see the cast of characters they were up against"—Rupert Murdoch at News Corporation, John Malone at TCI, Michael Armstrong at AT&T, Dick

Parsons at TimeWarner and all the rest—"and you'd wonder if the Robertses were up to the fight. Ralph is everybody's Jewish grandfather, pleasant and nice. They didn't seem to be locked and loaded like all those other cable CEOs.

"But they won! They're everywhere! Brian's now the biggest of big shots in the Philadelphia area!" he continued. "So un-Hollywood-like, so un-New York-like—[even though] Aileen and Brian do look like they belong in *Vanity Fair*."

Close to Home

Ralph had urged Brian to try a career in the investment business. But Brian didn't go back to Drexel, Burnham, or up to Wall Street, on his future breaks. "After that he spent his summers playing squash," according to Aronson. Once Brian was a newly minted Wharton grad, Ralph had him installed as an accountant in Trenton, New Jersey, the largest Comcast franchise within driving distance of home. The next year, 1982, he was sent all the way to Comcast's biggest subsidiary, Flint, Michigan, the General Motors factory town best known in recent years as the setting for left-wing director Michael Moore's comedy, *Roger and Me*, which sought to depict the "lighter" side of factory closings, home foreclosures, and corporate greed in his hometown.

Flint was the ideal cable TV market—a city full of workers with middle-class union paychecks (or unemployment benefits, when Americans periodically stopped buying cars), relatively few college graduates, and many dead-end jobs, worked by people who might be expected to want something to watch that went a bit beyond network television, and who would be willing to pay for it. Brian was made assistant manager in Flint. No more pole climbing, no more door-to-door sales, no more squash-playing summers, and no more investment-banking internships courtesy of Uncle Leon.

Flint had lately suffered another distinction within Comcast's off-brand cable empire. In 1981, not long before Brian arrived, Comcast was hit with the biggest fine levied to date by the FCC on any cable TV company—$20,000—for setting its Flint signal at frequencies that interfered with commercial jetliner communications. In trying to contact the FCC's Ohio air traffic control center, two airplanes flying five miles over Michigan raised static from Comcast satellite signals. The company had promised the government it wasn't using those frequencies, which were reserved for planes. Comcast showed "shocking disregard" not just for rules but also for the "safety-of-life concern that is the basis of the rules," the FCC determined in levying the fine. Comcast officials shrugged it off; after all, nobody had actually been hurt.[12]

There were no fines on Brian's watch. The next year he was back in Trenton, close to home for good. Within a year, he was the number-two executive in Comcast's eastern region; before that year was over he was second-in-command for all Comcast's East Coast operations; and by the end of the 1980s, he had joined his father on the Comcast board and served as executive vice president, in his father's inner circle. Brian joined the Comcast board in 1987, the year of the Westinghouse deal;[13] participated in the Storer negotiations and led Comcast's first relatively modest move into the cell phone business; and mixed with John Malone, Ted Turner, and others in the cable elite. A director of C-SPAN, Brian was in position to observe media policymakers in Washington as closely as he cared to; a director of Cable Laboratories, he was also situated to influence the course of cable-equipment research as media and software companies groped their way toward the Internet.

In February 1990 Ralph announced that his board had elected Brian president, second in command only to Ralph. "Brian is well prepared both technically and intellectually for his new role as president. He is highly regarded by leaders in the cable industry as well as the financial institutions which support us," his father affirmed.[14]

By the time of Brian's elevation he'd already been married for two years to the former Aileen Kennedy, a banker; their daughter Sarah had been born, her name recalling Ralph's mother. Brian and Aileen had met at a formal fundraiser for Philadelphia's Franklin Institute science museum. "I used the old 'Haven't we met?' line," according to Brian. "Well, I thought we had met."[15] If Comcast somehow survives the increasing turmoil of the media business as a family-run company, one or more of their three children may someday decide Comcast's destiny as he has. Brian holds his commanding block of Comcast voting stock in his name, and that of his heirs.

9
CONTROLLING
INTERESTS

"Ten Times as Rich"

The country path Ralph Roberts took to Mississippi in the 1960s grew into the Comcast superhighway that ran over AT&T and chased after Disney forty years later. Halfway down that road, Ralph Roberts could have retired rich, his frugal reputation intact. What drove Ralph and Comcast, instead, to the sneaking, opportunistic, overreaching monster deals that made Comcast tops in its business, and which would have done credit, in substance if not in style, to Standard Oil's John D. Rockefeller or Microsoft's William H. Gates?

One day in 1983, Julian Brodsky went in to see the boss, inspired by a plan for bringing the Comcast story to a profitable end.

"There is a strange thing at work in the land," Brodsky told Ralph Roberts. "Junk bonds. Leveraged buyouts." After forty years in business, twenty years in cable, they were none of them getting any younger. Their partner Dan Aaron, hardworking, humorous, and efficient, had cut back his workload to fight Parkinson's disease. Maybe now, with business conditions much improved and media companies in growing demand, it was time to think about selling Comcast. They could start by taking the company private, selling high-interest bonds, which were available as never before, Brodsky suggested. They could sell some assets, maybe buy some better ones; slice expenses, trim the debt, and prepare the works for resale in a few years at a fantastic profit, given the way all wired entertainment businesses were starting to take off that year, in 1983.

"The end result," Brodsky told Roberts, "will be that you and I will be ten times as rich as under any other scenario."[1]

Brodsky had been studying the work of John Werner Kluge, a German immigrant who had put together the nation's biggest independent chain of radio and TV stations. At age sixty-nine, Kluge had just figured out how to become a billionaire, using a combination of his life's work and other peoples' money. He had offered to buy out Metromedia's public shareholders with the new high-risk, high-yield debt. If he could tighten Metromedia's budget, pay off some of its obligations, and hold on through the next couple of years of media price inflation, Kluge hoped to be able to sell his company for a fortune. Of course, that depended on Kluge's being smarter or luckier than his investors. (He was: Kluge went on to pocket over $100 million directly from this deal, while simultaneously gaining control of Metromedia, which he sold to Rupert Murdoch for more than $1 billion three years later to form the basis of the fledgling Fox TV network.)

Roberts, the perfect listener, weighed each point. And let Brodsky down gently but firmly. "He says he appreciated it, but, 'Just keep doing what you're doing,'" Brodsky recounted, laughing at the memory. "He says he'd rather not leverage the company that much. 'Just keep the balance sheet so we all sleep at night.'"

All for Brian

But Roberts wasn't suggesting his company remain mired in the old ways. Far from it. It wasn't that Brodsky's vision went too far. It was that Roberts had no intention of giving up control to anyone, even at a rich price. Roberts had a vision of his own, and it went further than any investment banker's get-rich deal making. Ralph hadn't set out to be a media mogul; unlike men such as Benton and Shapp, who had claimed to see business as a means to social ends, Ralph had no great or world-improving personal agenda to project over his spaghetti network of wires, beyond the imperative to make it larger and more profitable.

He could have sold the business he'd cobbled together for a few hundred million in the early 1980s, or a few billion in the early 1990s, or a few tens of billions toward the end of the 1990s. Though he owned only a small part of the company he had founded, by the end of that period he could certainly have accepted a comfortable payment in exchange for ceding his disproportionate power over the company; it would have been enough to ensure his grandchildren never had to work, enough to dwarf every fortune his wife's old-money family had ever made, enough to carve his name on every university, hospital, museum, and concert hall in his adopted Philadelphia, with plenty left over for national and world causes. Sell is what Ralph's contemporaries mostly did. "The guys who really made money in this business," Brodsky said, "guys like [Gerry] Lenfest and [Amos] Hostetter [of Continental Cable]—they never went public. They kept on betting the farm, they took enormous risks, and they got the biggest paydays" when they chose or were forced to sell. But Ralph "never considered it."

Ralph Roberts had two powerful reasons for staying in the cable business instead of selling. One was the hope of greater profits through control of the new digital media, backed by easy finance, agreeable regulation and canny deal making—a vision already common, by the early 1980s, in Silicon Valley, on Wall Street, in Hollywood, across the cable business, and even in Washington policy circles. The other was his son. Brian was now ready to dive into the business, Wharton degree in hand, and fulfill his childhood offer to personally make Ralph the founder of a dynasty.

Not long after the Kluge interview, Ralph told Brodsky, "'I think I'd like to build a giant public company and hand Brian a nice thing.'"

Family Men

Digital, deals, and Brian—they were Comcast's future and Ralph's own crown. His son's arrival as partner and successor ended the Roberts family wanderings of the past three generations, in which Ralph, his father, and grandfather had each adopted a new profession, a new home, even a new name, as they moved from Russia to New York to Philadelphia.

This new order was much closer to the pattern of his wife's family, the Philadelphia Fleishers, who on arriving in America when it was still young had established a family seat and a capital base that grew and prospered successively in each branch and every generation. To be sure, the Fleishers, with their connections, the Sunsteins and Gerstleys, like most grand old American families, had eventually diffused across the landscape, selling the block-long factory and the downtown mansions, closing the firms that bore their names, withdrawing from Philadelphia public life and into the rich private spheres of inherited money (which tends not to be so prolific as self-made money) and professional success that circle the world. It was Ralph Roberts, who had relied on Fleisher money and Fleisher connections for the first twenty years of his career, who breathed new life into that old tradition and gave his wife's blood-line, in Brian, its most spectacular business success, right in Philadel-phia, the scene of its earlier triumphs.

Many captains of industry have dreamed of leaving Junior the works, but it rarely happens. Junior isn't interested; or he doesn't share Dad's talents, or the deep need that drove Dad to develop the mastery over resources, people, and ideas that makes it possible to build and run a giant corporation. The retirement funds and professional investors who control most public companies can't stomach too high a level of nepotism. Under modern, scientific management, can Junior really hap-pen to be the best man for the job? Of course, at Comcast, Ralph had in-sisted, even in his company's darkest days, on controlling its voting stock; when the time came, he could name his own successor and thumb his nose at Wall Street, if it came to that. Rupert Murdoch had done it, putting his sons in charge of News Corporation, and got away with it, ig-noring investors' outrage.

Yet even Comcast critics agree Brian came to the job prepared; ad-mirers say that was clear from the outset. Ed Breen, who rose from cable salesman to chairman at General Instruments (he now heads giant Tyco International), noticed the phenomenon from the start: "When Brian came into the company you could sense he wanted to be one of the big players." With Brian as with his dad, "you could sense [Comcast] was their life," Breen said. "They were not going to sell out. They were very methodical in their acquisitions, in their operations, in all they did."[2]

Without Brian's interest—and Ralph's intention to reward it—Com-cast would not likely have survived to menace Disney and AT&T. Yet pas-sion alone doesn't win empires. You need more than resources. You also need a program of conquest.

The Space Shuttle and the Titanic

"One minute we're about to blast off like the space shuttle. The next, we're sinking like the Titanic," Ralph Roberts complained to a group of his peers in the fall of 1984.[3]

He was complaining—not about the vagaries of the cable business, a fact of life to his knowing audience—but about the way business journalists write about it: "Over the years, our success and our failures have tended to be greatly exaggerated in the press." Ralph was referring, most immediately, to a *New York Times* article that had questioned cable's prospects, given its boom-and-bust history and the inflated prices cable assets had commanded in recent deals. Roberts wanted no room for such doubts: "Growth has not slowed." Instead, premium movie services like HBO, Showtime, and Movie Channel, with their menu of sex and violence unavailable on broadcast TV, had helped cable grow its customers sevenfold, to 27 million households, since the first cable satellites were launched in the mid-1970s. Never mind that actual, bankable profits remained in chronically short supply; like old-time canal-builders, railroad promoters, real estate speculators, and growth-stock gurus, cable has always directed attention toward how much it sells, not how much it keeps.

But Roberts saw a bogey on the horizon: the threat of real competition from the unwieldy satellite dishes that were already starting to offer an alternative to cable. And he urged his colleagues to embrace new scramblers, digital transmission systems, even service boycotts, if that's what it took to limit access to favored programs and exact a suitable toll from consumers.

"We will be de-emphasizing those programs that are indiscriminately carried by every other delivery service," Roberts warned. He urged his colleagues to prevent "the expansion of service from cable to other forms" of media.

"Let me make it clear. We don't mind competing for pathways to the American home," he said. "But I do not believe we should be competing against the very entertainment services we introduced, nurtured and helped establish." Comcast and its peers had "spent millions of dollars to establish brand identities in our communities for HBO, Showtime, Movie Channel," for "ESPN, CNN, MTV." So cable should also enjoy "a right to exclusivity for that programming."

Here was clear hypocrisy. Hadn't Roberts and his contemporaries grown parasitically, by snatching broadcasters' signals out of the rural air? Free signals had made cable possible; but since 1975 the industry had learned to pay for HBO and other wholesale programming it could retail for a profit, and it was determined to pull the ladder it had climbed to reach its newly lucrative position out of the reach of businesses and individuals who had hoped to find new and cheaper ways to sell the same programs cable carried.

The speech was seminal. Ralph Roberts had grasped early that cable had passed from the stage of charging for a free product, to finding ways to make sure no one else sold what cable controlled. Others thought as much and more, but Roberts put it into clear and unsentimental prose; his measured confidence in the rightness of his case helped make the industry's moves toward digital control respectable. "Ralph gave the industry a whole bunch of guidance. He took the industry into a new financial framework, a new regulatory framework," and new technologies, said Frank Drendel, then a General Instruments executive and now chairman of Commscope, the nation's largest TV cable manufacturer.[4]

Cable started at a disadvantage with other utilities, Drendel noted. For example, "The telephone industry was built around a switch. You charged for the time [a customer] spent connected. But over a network like television, you didn't have any way to charge for connected time. . . . Ralph saw the new services coming. He saw we had to control programming, both owning it, and distributing it."

Ralph's deal making, Brodsky's fundraising—impressing Breen and Drendel and their cable-equipment colleagues to go digital—it all served Ralph Roberts's master plan to put customers ever more tightly under the control of their cable providers, led by Comcast, and, in time, by Brian.

LIVING COLOR

Bang-Bang vs. Boom-Boom

Like the Michigan city where Brian had previously worked, Trenton was a fading factory town; it was also the worn home of New Jersey's state government. In 1983, when Comcast ranked as the nation's seventeenth-largest cable operator with 300,000 customers, Trenton was one of the company's largest markets. And it was near enough to Bala Cynwyd that Ralph and his team could stay close. The youthful manager struggled through familiar cable crises, like the time one of his crews was blamed for knocking out power to much of the city. But, like his dad, Brian kept an eye out for opportunities.

One April morning, Brian Roberts joined Trenton's mayor, Arthur Holland, and the aged boxing champion, Jersey Joe Wolcott, for a press event to boost Comcast's planned premium coverage of a championship bout in a political hotspot halfway around the world. They gathered at Comcast's Trenton office to honor a twenty-two-year-old lightweight contender, Kenny Bogner Jr., "a man of few words and quick fists," as Comcast's publicist put it, the local favorite in the planned South African show.[1]

Bogner was scheduled to challenge the reigning champ, Ray "Boom-Boom" Mancini, famed for hitting a previous challenger so hard the man died. Mancini and Bogner, who had taken to billing himself as "Bang-Bang," were double-billed with a middleweight title defense by Roberto

Duran. The card was cobbled together by American boxing promoter Robert Arum and his partner, Solomon Kerzner, who owned the site where the match was to be held: a hotel in Sun City, South Africa's answer to Las Vegas. Sun City was a white-run luxury resort in the poor rural "bantustan" of Bophuthatswana, which was both a reservation for Africans refused citizenship in apartheid South Africa, and a "nation" recognized only by that country's white regime. Even traditionalist South African whites disapproved of the resort; it was a haven for gambling and other vices, "Sin City" to the straitlaced elders of the nation's conservative Christian elite.

But money follows vice. Arum and Kerzner had guaranteed champ Mancini half a million dollars for showing up; even youthful challenger Bogner was guaranteed $200,000. The show was to be opened by the crooning of that famous New Jersey expatriate and previous boycott-buster, Frank Sinatra, enabling Comcast to bill the fight as "The Chairman and The Champs."

New York City refused permission to simulcast the fight at Shea Stadium. "It would be wrong for the city to enter a voluntary commercial relationship to promote an event taking place in South Africa," the parks commissioner told the *New York Times*.[2] One can imagine Arum and his fellow promoters shrugging; they had other screens, thanks to people like Comcast.

The Trenton gathering was timed for the eve of Bogner's planned departure from Newark Airport to South Africa. Comcast didn't exactly spring for a glitzy promotion. Bogner delivered the traditional challenge against the absent champ. Brian gave Bogner a going-away cake. Bogner's parents, trainer, and girlfriend cut the cake and served it to Brian and to other Comcast executives who'd driven up from Bala Cynwyd for the occasion.

What was Brian's interest in Bophuthatswana? It wasn't that he wished to promote apartheid; it wasn't just that he might identify with Bogner as a fellow member of the small, loose, youthful fraternity of internationally ranked Jewish athletes. It was that Comcast promised to televise the fight and a string of lesser bouts over two days for $39.95— one of its first forays into lucrative pay-per-view TV. The fights appealed to die-hard fans: Subscribers would have to stop by their local Comcast office and buy a "palm-sized TV decoder." But Brian was ready to innovate; other cable companies had been running similar events, with increasing success, since the Frazier–Ali "Thrilla in Manila" back in 1975.

Bogner flew off to meet Mancini; but after all that, the fight never happened. Mancini broke his collarbone in training; he never did face Bogner, and Comcast's experiment in fee-for-service TV was deferred, for a time.[3]

Urban Warriors

What made Comcast's willingness to deal with apartheid South Africa look especially cynical is that Comcast was simultaneously cozying up to African Americans in a place where blacks did enjoy political clout: Philadelphia. South of Trenton, Comcast was barely a factor in its hometown market. Of Philadelphia's three hundred suburban towns, only wealthy Lower Merion, where Comcast was based, and Willow Grove north of the city had chosen Roberts's company. Gerry Lenfest and other local operators would dominate the market until the day Comcast decided to buy them out.

After mayors Tate and Rizzo had failed to deliver cable TV, William J. Green, son of a Philadelphia Democratic boss and U.S. congressman, vowed to capitalize on the couch-potato vote by awarding new franchises.[4] Ten would-be cable providers hired lobbyists and ad agencies after Green took office in 1980. At first, the mayor tried to discard the neighborhood-franchise approach in favor of a single citywide contract. City council, elected mostly by neighborhood districts, resisted. In 1982 the two sides compromised, slicing the city into four territories where members could enjoy leverage that amounted to veto power. Green had to preside over a polarized coalition of black and white Democrats who had only begun to resume working together after Rizzo, tough-guy ex-police commissioner, had driven African American leaders to form a temporary party of their own. To preserve Democratic unity, City Hall forced each cable applicant to come up with "a plan that would allow local minority and female ownership participation in the system."

Comcast, though publicly traded, was controlled by one person— Ralph Roberts—who was neither black nor female. To get around that obstacle, Comcast hired the Butcher investment firm, which had taken Comcast public, along with smaller, black-owned Pryor, Govan, Counts & Company, to recruit 500 black or female investors. Comcast delivered the list of buyers to City Hall, where councilmen from black districts could read for themselves how many of their better-off constituents were on Roberts's team. To reassure urban investors who might recall the price collapse that had followed Comcast's original stock sale, Comcast even offered a special five-year, money-back guarantee.

What the arrangement didn't give the black investors was actual control: The shares carried no vote. By picking many small investors instead of a single partner who might want to share authority, Ralph Roberts ensured once more that power over Comcast would remain firmly in his own hands.

Two Sets of Books

Ralph and Brian Roberts excel at tailoring their messages to the audience at hand, at giving the audience what it wants to hear. In meetings with investors and groups like the Philadelphia Cable Club in the early 1980s, Ralph Roberts said his company kept 30 cents of every dollar it collected after deducting operating expenses, and that its profits had risen steadily since the company went public ten years earlier.[5] The company told Uncle Sam a different story. In its reports to the Securities and Exchange Commission, Comcast showed that, after paying its hungry bankers and bondholders, the company was making just 10 cents on the dollar in 1982.[6]

But who cared, if you could turn even a modest profit while grabbing market share? "Cable companies are buying each other up. There will only be a handful left, and we want to be among them," Julian Brodsky told the *New York Times* after buying a Baltimore-area system in 1983.[7]

Comcast wasn't just looking for more customers; now that cable was starting to attract advertisers, the company was also going beyond its old cheap-to-wire strategy and looking for more affluent customers. In Michigan, Comcast expanded from shrinking Flint to a growing string of Detroit's wealthy northern lakeshore suburbs, thanks to a timely divestiture by rival Cox Communications, which needed to sell cable holdings in order to buy a coveted Detroit TV station. The deals left Comcast with almost 1 percent of the nation's cable households, which, thanks to satellite transmission, HBO, and better equipment, had tripled to more than 30 million homes since the mid-1970s. The industry leaders, John Malone's TCI and Time, each held around 7 percent of the market. Rival Ted Turner's Turner Broadcasting claimed a profit for some years by signing a network of local cable companies to carry its proprietary programs, including its Atlanta "superchannel," and leveraging that network to collect as much as half the nation's cable advertising in the mid-1980s.

As in previous cable booms, however, most of the big players—including TCI and Time—were spending more than they were bringing in. To curb its million-dollar-a-week losses, one of the biggest, Warner/Amex, a joint venture of Warner movie studios and the American Express travel conglomerate, hired Pennsylvania Republican Drew Lewis, formerly Ronald Reagan's transportation secretary, to execute a turnaround. As part of his cost cutting, Lewis promptly canceled agreements the company had made to build studios and offer other perks in a number of cities where Warner held the local cable monopoly.

Whatever his management skills, Lewis's political connections were undeniable.[8] With friends like Lewis, cable won a signal victory in Washington: In 1984, the most Republican Congress since the 1950s passed a Communications Policy Act that stripped city councils and state public utility regulators of their right to set cable rates.[9] The new policy, effective after 1986, would move cable politics from late-night, local-government meetings to Washington, D.C.—the big-time at last.

What You Can't Get

To fund their expanding operations, operators like Malone and Turner eventually turned to Drexel, Burnham, Lambert, the firm where Brian Roberts had interned. The firm had been transformed from a Philadelphia old-money brokerage into a bicoastal powerhouse, the nation's best-known seller of high-risk, high-yield junk bonds. In 1984, Drexel helped the New York buyout firm of Kolberg Kravis Roberts beat Comcast's audacious $2.5 billion bid for Miami-based Storer Communications. Storer said it chose KKR because it believed Milken's junk bonds were more reliable than Comcast's bid, which was funded by a group of banks, led not by the Philadelphians but by the Bank of Montreal, plus $1.2 billion in guarantees from Merrill Lynch, which depended on a scheme to raise funds by selling Storer's broadcast stations.

It wasn't a complete loss for Comcast. Treasurer Bernard Gallagher, like Aesop's fox, scorned what he couldn't get, boasting his company was one of the industry's most profitable because it refused to overpay for its many acquisitions. Ralph Roberts again bragged the company had never had a down quarter since going public a decade earlier—which was true, but only if you counted total profits for the growing company, not profits as a percentage of sales, which tended to plunge after every acquisition. Roberts's message got great play. Comcast was "financially superior," said the *Washington Post*.[10] A "little-known" company, but "one of the best-run," and "hardly spendthrift," said the *New York Times*.[11] "Cautious" and "prudent" and "methodical," said trade journal *USA Broadcasting*.[12] Roberts was getting high marks for patience, which even Comcast's first-ever reported loss later that year didn't dispel. Storer could wait.

Growing and financing the company took up much of Comcast's top executives' time. Brian was vice president for operations; Ralph still handled strategy and deal making, while handing the key areas of programming, marketing, and acquisition management to veteran cable executives like Robert Classen, lured from Canada and appointed president of Comcast's cable unit. Classen was the first of a series of outside executives who would mind the store as Brian came into his own.

When American Mutual Funds started its popular New Economy Fund in 1983, with the express intention of taking advantage of those stocks that would ride the new technology and telecommunications boom, it listed both Malone's TCI and Comcast among its top holdings—if only because, with their weak stock prices, it looked as though someone might come along and buy them.

COMCAST COUNTRY

"Sinister and Diabolical"

Philadelphia City Council President Joseph Coleman was outraged. After twenty years of bickering and false starts, 1984 was the year he had planned to deliver two million Philadelphians to the finest, most politically balanced consortium of cable TV monopolists his Democratic majority could deliver. But Mayor Bill Green was in the way.[1] The mayor, a Democrat, too, saw things differently. He had rejected council's cable choices. His cable consultant had approved his alternative cable picks. They also had good party connections. Why should he defer to Coleman? The mayor and the council could not agree; and this was too bad for Comcast, because Ralph Roberts's company was playing both sides.

Since 1964, frustrated newspaper publishers, broadcasters, national corporations, and homegrown partnerships had kept local lobbyists happily employed chasing four mayors and a circus of councilmen in hopes of winning Philadelphia's cable concession. The city went through five separate application cycles, and awarded at least ten franchises over the years. Most expired before anyone got wired. Only Times Mirror Corporation had built anything—a small South Philadelphia system, which, though its contract had run out, kept operating in the vain hope of renewal. Tons of cable applications lined office floors at Philadelphia City Hall; passionate hearings echoed under the oil portraits of forgotten politicians; and Philadelphians had to be content with their old rabbit-ear antennas, G-edited UHF late-night movies, and away-game sports on free broadcast TV. Smaller cities and suburbs got cable, and still Philadelphia made do without. It was as if the city franchise was an employment program for lawyers and lobbyists, giving their political patrons little incentive to end the happy show by actually making service available.

Finally, in 1983, with elections looming, and city voters demanding the extra pro games, sports, and R-rated movies that viewers in most of the country now took for granted, council president Coleman and his backers had finally weeded out dozens of applicants and announced their choices—a list that seemed to mix regional obscurity with cold political calculation. But even that decision wasn't final. In August, council granted black-majority North and West Philadelphia to James N. Wade, an African American businessman and former aide to cable pioneer and later Pennsylvania governor, Milton Jerrold Shapp; the city's diverse northwest to Delaware trucking millionaire John Rollins; Center City and South Philly to Peter Bordes of Princeton, an Ivy League–educated radio and weekly-newspaper baron. The biggest, richest, and whitest territory, Northeast Philadelphia, was set aside for Ralph Roberts's Comcast.

Never mind that Rollins was plagued by labor troubles and service complaints so severe that Wilmington was trying to induce someone— anyone—to build a competing system; or that Bordes had been added to the team at the last minute after a previous favorite dropped out. This was council's team, and Philadelphia City Council was a group whose members valued loyalty.

The trouble was, Mayor William Green had settled on a rival slate of franchisees, complete with an imported minority politician, former Manhattan borough president turned radio station owner, Percy Sutton. (Times Mirror, the only company that had actually built any cable in Philadelphia, failed to make either team.) So what if the mayor's slate happened to include one or two of his old law partners? Battling with council, Green tried to seize the moral high ground. He told the *New York Times* his team was selected "strictly on the merits," implying, as his aides whispered, that council's picks were the ones who had been politically compromised.[2]

That outraged Coleman. "He is trying to con the public into believing that he and he alone is the only honest and decent and dedicated public servant in the city," the chairman fumed. He called the mayor's implication "the most sinister and the most diabolically concocted lie" in the city's cable history.

So no deal. That was a setback for Comcast—which stood to win no matter what the city decided—but not if the city kept deferring a decision.

Elsewhere, things weren't so bleak. Comcast bought its way into suburban Baltimore with the 1983 purchase of Calvert Telecommunications for $118 million in cash and bonds. Comcast cut back-office jobs, jacked up monthly bills, and pushed new cable subscriptions hard. Soon it was boasting of new subcribers and higher profits thanks to Baltimore, presenting Calvert to investors as an example of what Comcast could do elsewhere if they gave it the chance.

But Philadelphia would have to wait a little longer.

Let the Feds Do It

One of the powers guarded by American towns and cities is the right to lay pipes, poles, tracks, and cables through public streets and private property, and to regulate the services and prices charged by those "public" utilities, even when the managers are private contractors. Municipal contracts—water, electric, sewer, transportation, and eventually cable— have a rich history of corruption; but they have also given local residents a focus for venting their anger at poor service and price gouging, and even for goading their elected representatives to replace an incompetent or arrogant operator with someone more efficient and responsive.

But once cable had worked its way into urban markets, the industry mobilized to reduce the hassle of keeping local officials happy. In 1984 President Reagan signed a law that took the power to set cable rates away from towns and states and vested it in the Federal Communications Commission. It was a historic decision, and a cynical one. Reaganites were against central government power, except when it allowed favored industries to escape local regulation. To be sure, broadcasters had often used the FCC to restrict cable. But Reagan's appointees, as professional regulation skeptics, could be counted on to use that power sparingly.

The 1984 act didn't take effect until the beginning of 1987. It was followed, predictably, by a wave of price hikes, which encouraged investment in cable companies. By the end of the decade, cable had become so lucrative—and cable customers so angry—that Washington was echoing calls to "re-regulate" the industry.[3] One of the loudest advocates, Sen. Albert Gore of Tennessee, would soon be in a position to help make that happen.

The Biggest Piece

W. Wilson Goode, elected Philadelphia's first black mayor at the end of 1983, saw the job as a stepping-stone to something bigger: maybe the chairmanship of the nation's premier corporation, General Electric Company, which had been the biggest employer in his Southwest Philadelphia neighborhood before Chairman "Neutron Jack" Welch turned its chain of world-class factories into a global finance company. The buttoned-down mayor inherited his predecessor's rocky relationship with the truculent city council. Just before Christmas 1984, Goode finally yielded the power Green had refused to give up, and council approved the team it had selected under the previous mayor—or most of it: Action was deferred on Rollins because his company happened to be on strike, one of the few sure-fire deal-killers in a labor town like Philly. Comcast had won the city's choicest cut.

Northeast Philadelphia was a big place, by previous Comcast standards—more than three times as many customers as Trenton, and more than twice as many as Flint, which had been the company's biggest system. In addition, most of the houses were so close together that they shared walls, though all the digging and wiring needed to hook up the area's Irish, Italian, Jewish, German, and Polish neighborhoods would depress profits for a little while, Roberts warned investors. Some strings were attached to the deal: Comcast agreed to abandon its suburban headquarters and move its modest head office downtown. The old office had expanded from a sparsely furnished room to a crowded set of suites with paintings on loan from a downtown gallery operated by Suzanne Fleisher Roberts's niece (a cost-cutting move that occasionally backfired when an angry artist arrived to seize a painting for which he alleged he hadn't been paid).[4]

Comcast moved into Meridian Bank's Center City tower in the middle of a long period when old Philadelphia companies were closing, selling, or just moving out, leaving the downtown littered with half-empty towers and outmoded signs that gave the place the feel of a corporate museum. Comcast kept growing; when asked, Roberts and his team, which had made so many friends during the long campaign for a franchise, were careful to say nice things about their hometown. City council would repay Roberts's fidelity richly in the years to come.

In winning Philadelphia, Roberts had survived whole generations of competitors. Northeast Philly had originally been slated for division between the city's dominant newspapers, the morning *Inquirer* and the afternoon *Bulletin*; but, in the cable business, Comcast had outlasted them both. The company also outlasted Mayor Goode's larger ambitions, which faded past recall when his police force accidentally incinerated an armed back-to-nature cult and a block of pleasant West Philly rowhomes in 1985. TV caught the bombing and the resulting inferno and reran the images around the world, ensuring that the sober and ambitious mayor would end his working life as a Baptist deacon, not a corporate CEO; though that didn't stop him from being elected once more as mayor of Philadelphia.

Cable carried pictures of Philadelphia burning all across America. The city's reputation suffered, but Comcast kept growing. Philadelphia was now its loyal home market, the company's partner in consolidating its power and resisting competitors.

"What's Next?"

Pittsburgh is even less likely as a media center than Philadelphia. Yet that's where George Westinghouse put his generator factory back when electricity was new, and that's where his company started the commercial broadcasting business with KDKA Radio in 1920; the station grew into Group W, one of the big radio and TV broadcasting chains long before Westinghouse bought CBS in 1995 and the Infinity radio stations a year later. Group W was perhaps the last champion of regional feature television against the dominance of Manhattan and Hollywood; its popular Mike Douglas Show and portions of its Evening Magazine show were produced in places like Cleveland and Philadelphia. But in 1980, Westinghouse bought into the national cable rush, purchasing Irving Kahn's old TelePrompTer cable networks and more than a million customers, plus a handful of stations, for what was then regarded as the extreme sum of $780 million in stock and assumed debts. That made Westinghouse the rival of Time as America's top cable company. Time itself had expanded from HBO into cable systems. American Express and the Warner movie studios had set up yet another cable network. The big bucks these corporations spent on acquisitions drove up the value of the remaining independent cable operators, and rumors had Cox, Viacom, and Comcast ready to sell if prices stayed high.

"It is a sign of the cable frenzy that has overtaken the American entertainment industry that some are suggesting [Westinghouse's] pricey bid might be topped," wrote Britain's *Economist*, in a masterpiece of backward-looking business journalism.[5] Kahn told the magazine that fiber-optic cable and computers would soon be sending, not just TV shows, but also "newspaper stories, travel guides, shopping catalogues and electronic mail" to millions of homes. Already the cable visionaries were linking their wires to what wasn't yet called the Internet. Kahn thought it would all arrive by 1985.

But like other big corporations that sampled cable, Westinghouse didn't have the patience to stay in the game through lean days as well as fat. Unlike Comcast, whose founder still controlled the voting stock, Westinghouse had to answer to outside investors, who had little patience for expensive campaigns that failed to translate quickly into higher profits. In 1986, after frenzied negotiations, Westinghouse's cable systems were split among Time, John Malone's TCI, and three smaller regional companies, including Comcast. At that moment, and for some years to come, Comcast was overshadowed by its larger partners in the deal; TCI and TimeWarner built big diversified entertainment companies, while Westinghouse kept shedding factories and acquiring TV and radio properties, dropped its old name and took on that of CBS, and was finally sold to Sumner Redstone's Viacom, for $40 billion, in 1999. Still, by buying a piece of Westinghouse, Comcast had not just maintained its independence, but also established itself as a cable acquirer on a national scale; young Brian Roberts and his advisers had gained deal-making experience they would put to greater use.

The Westinghouse arrangement also provided Brian Roberts with his first working exposure to another bright young man, Steven Rattner, who was then an assistant to one of Ralph's favorite investment bankers, Felix Rohatyn of Lazard, Freres & Company in New York. Rattner, a former standout New York Times reporter, who had made the leap to Wall Street and become, along the way, a leading Democratic Party financier, has been at Brian's side through key points of his career. With Rohatyn, Rattner helped Comcast win control of QVC. In 2000, he quit Lazard and joined a group of colleagues in founding a new firm, Quadrangle Group, which counted Brian and other blue-chip media executives among its initial investors. And in the winter of 2003–2004, Rattner would advise Brian's ill-fated bid for Disney.

The Westinghouse deal cost Comcast $1.6 billion—a fraction of the value of its later takeovers, but a great leap for the company at the time. It doubled Comcast's customer base and ranked it for the first time among the larger national operators. By that point, no one was thinking about Comcast as a seller; it was instead becoming a relentless buyer of cable assets. "Our intention," Ralph Roberts told the weekly *Broadcasting* that summer, "is to continue to grow."[6]

Two days later he assembled his top aides and asked them, "What's gonna be next?"[7]

SHOPPING

Song and Dance

Tucker Roberts, son of Brian, six months shy of his bar mitzvah, stood with his fellow singers of the Keystone State Boys' Choir, looked over at his smiling grandfather Ralph, and belted out this tribute:

> Who can make subscribers
> Pay for what they see
> And have them watch it on a set
> They bought from QVC?
> The cable man can . . .

This cheery commercial parody of a sweet 60s show tune,[1] sung by innocents, was perfectly appropriate to QVC, a channel that subverts traditional television: its cable viewers pay to watch advertisers, who are the show; then pay again, sending their money directly to the channel, which replaces the traditional store. QVC owed its existence and its great success to Ralph Roberts, the day's honoree.

In the audience sat the outrageous TV personalities, not the half-dressed actors from Comcast's E! The Entertainment Channel, but the men behind the little screen, the famous dealmakers and visionaries, who had grown old watching Roberts's power expand at their expense:

Barry Diller, whose ambitions for himself and QVC Ralph and Brian had crushed; Ted Turner, whose epic bailout they had joined with such a determined lack of enthusiasm; Chuck Dolan, from Cablevision in New York. Suzanne Roberts was also there, by video, recalling how she thought her husband was "absolutely crazy" for having gotten involved in cable TV the way he did long years before. They had gathered in June, 2003, in Washington, D.C., to honor Ralph with the National Television Academy's Trustees' Award—an honorary Emmy, as it were—a prize previously given to President John F. Kennedy and TV anchor Walter Cronkite, men great masses of Americans trusted, or at least knew, or thought they did.[2]

In any tribute to Ralph, it was appropriate that QVC get its share of the credit. The world's biggest electronic retailer was the best investment Comcast had ever made, providing the greatest, steadiest, and fastest-rising share of the company's sales and profits. It wasn't Hollywood; it was better than Hollywood: efficient, bankable, always lucrative. QVC existed largely because Ralph had believed in it; Brian had embraced it, and made great predictions for its future, for a time. But that time was almost over. Brian and Comcast were moving on. Before the year ended, QVC would itself be sold like a great zirconia, clearing the way for the family company's next ambition, and serving, Brian hoped, as an example of how he might be trusted with even bigger properties outside his usual cable beat.[3]

Soft Sell

It made Joe Segel fidget. A millionaire marketer, onetime teenage prodigy at Wharton, founder of Franklin Mint collectibles, semiretired to Florida by the mid-1980s, he was intrigued by television's Home Shopping Network but put off by what he considered its smarmy, high-pressure approach to televised merchandising. Back in Philadelphia, Segel told Ralph Roberts he knew he could do a better job, and Roberts encouraged him, promising to help provide financial backing for the venture.

That's how, in 1986, Comcast became a charter investor in QVC—Quality, Value, Convenience, the home-shopping channel. Joined by Warren "Pete" Musser, who had sold Ralph his first cable system, Comcast made QVC its biggest and most successful attempt, prior to its Disney bid, to own TV channels instead of just retailing programs to viewers. The Musser–Comcast group raised $20 million, more than Comcast's annual profit in those days. Segel began what became a sprawling TV-studio complex ten minutes from Ralph's Brandywine Valley estate, on the outskirts of the old courthouse-and-college town of West Chester, about as far from Hollywood as America gets. Professional pitchwomen and company owners who presented products were told to stress information—what's special about the product, how to use it or show it off to advantage—instead of the hard sell. In time, QVC became a tourist attraction for busloads of QVC's loyal viewers, as well as visiting merchandisers, the people who paid Comcast to buy and to sell. The prospect of marrying QVC to a Web site especially excited Brian, and Comcast investors, during the Internet bubble of the late 1990s.

And why not? "The Roberts[es] describe themselves as opportunists rather than grand strategists. . . . They have fingers in several pies," wrote a British observer around that time.[4] In fact, there was never a time when Comcast was only a cable company. Every time Comcast bought another string of cable monopolies, Ralph Roberts seemed to hedge his bets with another investment outside its main business of U.S. cable. QVC followed closely on the Westinghouse cable purchase, which doubled Comcast's cable subscribers to two million. In 1988, as Ralph closed in on his third attempt to buy Storer Communications and another million viewers, Comcast entered the cell-phone business, purchasing American Cellular Network Corporation in New Jersey. Starting in 1984, the company had also invested in a British cable network. Comcast still owned the nation's biggest Muzak canned-music franchise, based in Texas (it was sold, to Muzak managers, in 1993). (Of course, Comcast didn't buy everything. Ralph declined to play a major role in the cable industry's collective bailout of near-bankrupt Ted Turner in 1987, buying only a nominal $5 million stake in Turner Communications, instead of the $100 million Comcast was offered. It was a decision the family later regretted, as Turner recovered handsomely. But Brian did serve on the board of Turner Communications, joining John Malone and other industry heavies, at age twenty-nine.)

Glamorous in a small, product-specific way when designers like Diane von Furstenberg arrived to shill their wares, QVC made its greatest impact on the accounting ledger, where it accounted for a third to half of Comcast's business. QVC's expanding sales enabled Brian Roberts to lure a genuine Hollywood magnate, Barry Diller, all the way to West Chester to run the channel in 1991. That put QVC on the map: Diller had worked at Paramount Pictures and ABC-TV; he had green-lighted shows ranging from *Star Trek* to *Die Hard* to *The Simpsons*. Later, he had put together Rupert Murdoch's Fox network with old John Kluge's Metromedia stations. Having parted company with Murdoch, taking with him over $100 million in what amounted to severance pay for his next venture, Diller had cast about for a network of his own, but found none available at the time. Flying to Philadelphia on his Gulfstream jet, he met with Brian and Ralph at that elegant chain hostel, the Four Seasons Hotel. A tour of the bustling QVC site, where he met the equally elegant von Furstenberg, clinched it for Diller.

Diller even brought his own capital to the deal: He committed $25 million to invest in the company. He obtained a helicopter, to fly from West Chester to Philadelphia and New York. For a while, his path and Comcast's ran together. Of course, Diller hadn't come all the way to West Chester just to stand in Brian Roberts's shadow. QVC was a cash machine that needed Diller's particular blend of panache, aggression, and mass-market taste, while providing him with a stable financial and business base for his goal: attracting enough additional capital to build his own empire and compete with Hollywood's biggest. But that set Diller squarely across Comcast's own ambitions for QVC. For Comcast was using Diller, too. And the Robertses wouldn't be pushed aside so easily.

"It's Shopping, and It's TV"

Diller's first target was naturally his old studio, Paramount Communications. But he faced a powerful and motivated rival: Viacom, the owner of MTV and Nickelodeon, whose boss, Sumner Redstone, was engaged in building an empire the old-fashioned way, from his own corporate base, one property at a time. The clash of fortunes and egos, with one of Hollywood's most venerable studios as the prize, played out in New York investment bankers' offices; in the business-friendly Delaware Court of Chancery in Wilmington, where corporate America traditionally settles its ownership disputes; and in boardrooms across the country. Redstone enlisted financing from BellSouth; another big regional phone company, New York's Nynex, backed Diller. Redstone found more cash from Blockbuster Entertainment (which his Viacom later bought); Diller secured the Cox family's cable and publishing business. Each bid finally topped $10 billion; Diller actually outbid Redstone by offering a quarter of a billion extra—but Redstone guaranteed his price against the changing value of his company's stock, which Diller wouldn't match. Paramount took Redstone's money, and his shareholders spent the next few years recovering from the cost. Diller issued a famous statement: "They won. We lost. Next."

Diller switched from movies to TV; if not Paramount, maybe CBS. He had targeted what had once been the prestigious "Tiffany network" on Rupert Murdoch's behalf when he cobbled together the Fox network and crammed it with fast-paced, youth-oriented soap operas. Now Diller saw faltering CBS as his chance: If it lacked Paramount's cinematic production capacity, it had the advantage of being available, as CBS's owners, New York's Tisch family, reached the conclusion that the network's solution was beyond them. With a motivated seller, Diller thought he had a deal.

But Diller's vision left little room for Comcast—and Ralph Roberts had other plans, which Diller wasn't part of. He assembled his dealmakers' brain trust, the same Philadelphia and New York hands who had worked on his biggest deals of the past decade. Like a powerful old judge in a courtroom of his own creation, Ralph Roberts settled back into a familiar role: listening while his loyal deputies and his highly paid servants pleaded and argued, cited and warned. Diller's ambition plus QVC's dire need for expansion capital had forced Comcast to a tough decision. Diller's CBS plan meant bringing in outside capital; that would likely mean new investors and reducing Comcast's role and its QVC profits. That didn't work for Ralph; by late 1993 he was weighing his New York advisors' counterattack: a proposal to borrow the $2.1 billion it would cost to grab control of its profitable but troublesome home-shopping channel.

Earlier that year, Comcast had approved another big deal, the $1.2 billion purchase of Maclean Hunter Limited's cable operations in Florida, New Jersey, and Michigan.[5] Should QVC take precedence now over other possible cable deals? Comcast was wrestling with the question that dogs every successful enterprise: how best to deploy its hoarded cash and hard-earned credit. When he'd chosen not to sell out at the beginning of the 1980s, Roberts had forced Comcast to grow. Now that America was mostly wired for cable, the business was running out of new customers. With the Democrats back in power, the cable companies had been forced to cut rates and felt pressure to keep them in line; if revenues and profits were to rise, Comcast had to have new products and find new markets, so Ralph's son and heir could inherit the "special" and growing company he envisioned, instead of a mere annuity from a solid, boring utility company.

Comcast had always sold other people's TV programs; was it time to shell out big bucks for its own? Roberts had put together the visionaries who spoke of leveraging Comcast's strengths and gaining control over its destiny, the accountants who worried about debt and dilution and manager distraction, the investment bankers who seemed to know everyone on Wall Street. They produced charts that showed rivers of projected cash flowing into deserts of all-too-real debt—and waited for the final silence, which only he, as boss and majority voting shareholder, could fill.

Roberts's trusted adviser, Felix Rohatyn, the New York financier and Democratic Party moneyman who once saved New York City from bankruptcy and who, with his younger protégé Steve Rattner, had helped Comcast swing its piece of the Westinghouse deal, put it on the line: "You can make it work as long as QVC grows 8 to 10 percent a year."

"Of course it will grow that much," said Roberts. "It's shopping, and it's TV."[6]

Nothing Personal

Diller was jetting back from a meeting with CBS in Los Angeles, thinking he'd won his audacious bid to turn his West Chester vacation into a permanent assignment as boss over the bi-coastal network. Diller's private jet brought him into the Teterboro, New Jersey, airport. Brian and Ralph Roberts were there, waiting for him. "The Robertses hand Diller a letter saying they're making their own bid for QVC, a staggering move that prompts CBS chairman Laurence A. Tisch to scrap the merger," the *Philadelphia Inquirer* reported.[7]

"I don't think this is personal at all," Roberts said, adding he hoped Diller would stick around. But Diller was finished with West Chester. (He later moved on to serve a stint at the original Home Shopping Network before serving as head of USA Interactive, a programming/Internet hybrid. Like other objects of the Robertses' subterranean maneuvering, he has claimed to bear them no ill will and continues to move in their media circles.)

Wall Street initially hated Comcast's QVC deal, as it does most acquisitions on the day they're announced, and Comcast stock suffered one of its periodic collapses.[8] But Comcast's two QVC investments, in 1986 and 1993, were deals that shareholders would have applauded, if they could have seen ahead. The company pumped a total of $250 million into QVC, and was rewarded with its largest source of sales and profits. And then in 2003, to help pay for the AT&T deal and build confidence for what turned out to be its Disney bid, Comcast sold its QVC shares for nearly $8 billion to John Malone's Liberty Media. For each dollar invested, Comcast grossed $32. Why? In 2000, amidst the online shopping frenzy, QVC "answered 130 million phone calls, shipped 80 million packages. That's an average of 2.4 packages every second, every day, all year long," Brian Roberts boasted. "If you take L.L. Bean, Lands' End, eToys, and Amazon and combine them, QVC shipped more."[9]

Brian and his staff had negotiated the QVC deal, but at the end of the day it was still Ralph's call. Comcast's deals followed a basic pattern: Brian would negotiate, then go to his father for advice, for blessing, for final approval. "There's a moment that comes in every deal when you have to decide whether to pull the trigger," Brian Roberts said later. "You've gone farther than you said you'd go and the financial guys have already bailed, convinced the deal no longer makes sense. It's a lonely time, but my dad never blinks. He's been there so many times, and he doesn't lose sight of the bigger picture."[10] It took until 1995 for Comcast to complete its QVC investment, with John Malone's TCI as a minority partner. By then, Comcast was buying yet another cable system with nearly a million customers, E. W. Scripps, plus the part of Storer it didn't already own.

Brian Roberts's decision eight years later to sell QVC at a huge mark-up confirmed, if anyone doubted, that he'd absorbed the buy-low, sell-high aptitude his father had shown ever since his Pioneer Belt & Suspender days. But it also underlined the challenges facing the company under Brian. For if QVC had become the most reliable part of Comcast, its sale meant the company would rely more than ever on cable-based services, and less on programming or alternative ventures—at least until the next big deal.

Air and Wire

Air is an enemy, if you're a company that sells signals delivered to customers by wire. Phone and cable wires, when linked to services people want, can be licensed, metered, billed, and, if necessary, shut off. This allows their owners to collect a steady stream of cash payments. Wireless signals that can be sent or broadcast through air give customers an alternative. They are a threat to companies that depend on wire. Phone companies, whose wires once gave them a monopoly on long-distance signals, have long been required by federal "open carrier" rules to share their lines with upstart competitors. Cable companies, by contrast, have resisted being forced to share. Cable gained at the expense of broadcast TV by offering to their fee-paying, wire-bound customers programs that free TV couldn't match. But by the late 1980s, improvements in satellite technology—the same satellite technology that gave cable its programming edge—threatened to bypass cable and end its monopoly power. In- creasingly, people got their TV from a satellite dish, a situation cable companies did their best to fight by controlling local sports program- ming and lobbying to preserve the right to keep their shows off the satel- lite companies' schedules.

But perhaps the biggest threats to cable's dominance—in the minds of executives like Brian Roberts, who attended technology executive gatherings such as the retreats held by Paul Allen, Microsoft cofounder and cable investor, in Sun Valley, Idaho, every summer—were wireless technologies in the personal communications device, or cell phone. Cel- lular threatened both cable and that older wire network, the nation's phone system. At a time when most Americans were still getting used to the idea they might put a phone in their car, media executives were growing obsessed with the promise, and threat, that cell phones, picked from a range of competitive providers, might become personal, au- tonomous computers, shipping pictures, shows, text, and all sorts of data around the country. Consumers would be their own masters; that liberation, if it came to pass, could make the entire, expensive wire net- works for cable and phone obsolete.

The phone companies faced the most direct threat—voice is much simpler than video. In self-defense, regional Bell companies like Bell At- lantic and Southwestern Bell bought or started their own wireless sys- tems. AT&T was attempting a national cell network, challenged by and then allied with entrepreneur Craig McCaw. As in the early days of elec- trical service, when Westinghouse and Edison ran competing, incompati- ble power grids down the same city streets, these companies used differ- ent frequencies, signals, and relay towers; subscribers of one company were often cut off from rival networks—and companies worked to keep it that way.

In 1988, Comcast bought into the game, taking over American Cellu- lar Network Corporation in New Jersey and Delaware. Even with two million customers, AmCell was too small to go it alone; Comcast promised AmCell President John Scarpa a measure of independence that would be unlikely if the company were rolled into a regional Bell or national network.

Hanging Up

Four years later, in 1992, Comcast bought out John Kluge's next empire, Metrophone of Philadelphia, for a cool $1.1 billion. Cellular "reminds us of cable 25 years ago," Brian Roberts said at the time. "This acquisition is a real strategic fit." Vice president Mark Coblitz affirmed the "natural connection between cellular and cable to deliver" personal communications systems.[11] "Welcome to Comcast Country" billboards sprouted wherever the phones could be used. Comcast now had more cell phone subscribers—7 million—than cable customers, and it wasn't done. In 1996 the company offered $650 million for a stake in OmComm, which was trying to put together another nationwide network (Nextel) to compete with AT&T.

Ralph Roberts had abandoned belts and men's accessories years before because of too much competition; he loved cable because, after the initial outlay, it was a lucrative cash business. But as the 1990s wore on and cellular companies began cutting prices to compete, Brian began to wonder if there could be too much competition to make money, even in such a dynamic industry as cellular communications.

In the end, Comcast Cellular proved a detour. In 1999, Comcast sold its cellular arm to SBC Communications for $1.7 billion, a disappointing return. By then it had also given up its British cable franchise, another disappointment: On that densely populated island, competing cable networks suddenly found themselves under assault from the air—not from cell phones, but from British Sky Broadcasting, the latest audacious project from the Australian-born, global newspaper and broadcasting tycoon, the man responsible for America's rising Fox Network: Rupert Murdoch.

SPORTS

Owners, Not Fans

Philadelphia is where satellite TV salesmen go to die. Comcast Cable provides exclusive coverage of the popular 76ers of the National Basketball Association and Flyers of the National Hockey League. Comcast owns these teams outright, along with the city's main indoor venues, the Wachovia Center and the Spectrum. It also dominates Phillies baseball games, transmitted, just like the winter sports, through Comcast SportsNet, which includes pre- and postgame shows. All sports, all day, all night. And all Philadelphia. Satellite dishes have just half the market share they enjoy in other big cities.[1]

Don't expect pity from the governor of Pennsylvania, Ed Rendell: As mayor, he gave would-be Comcast competitors the bum's rush; he also hosts a Comcast SportsNet TV show, giving his weekly opinion on the hometown NFL Eagles. (Comcast doesn't only shoot high when seeking commentators: In 2000, it picked a heavy-set security guard from the SportsNet office, Fred Bibbo, as its wrestling commentator. Bibbo won a regional Emmy award for his work in 2001.)[2]

Philadelphia should be glad to have one big sports provider, Comcast maintains. Comcast has linked the old Philadelphia-area Rollins, Storer, AT&T, Lenfest, Greater Media, and Adelphia systems with fiber, the easier to move sports, video, and Internet reliably to customers.

"Everything we do is up for inspection," Vice President Ed Pardini told *Cable World* in 2002.[3] "Our operations need to be pristine because the last thing we want is to have any kind of major customer service issues in our home town." Comcast enjoys similar domination in the Baltimore–Washington corridor. It owns pieces of sports networks in New England and the South. And it's made arrangements with Chicago's pro teams to take over sports programming there in the fall of 2004.

Pro sports and cable TV go together like long, skinny hot dogs and flat, expensive beer. Team sports, as Brian Roberts sees it—like few, if any, other local television shows—can attract a lucrative audience in competition with the aggressive, deep-pocketed national entertainment media. Sports have proved far more durable than locally produced entertainment programs, which have all but vanished from TV because they attract narrow, not mass, audiences, and usually look pretty lame compared to glitzy Hollywood productions. Local news is still highly competitive because nobody has yet figured out how to centralize the reporting of local crime, weather, shaggy-dog stories, and other provincial staples; though Comcast is trying.[4] But each pro game can be purchased exclusively by a single programmer, who can charge monopoly rates for advertising—and everything else, in the case of the growing number of programmers who have taken over pro teams, along with their big-city stadiums and lucrative merchandise sales.

It's becoming a dim memory, but, half a century ago, broadcast TV had liberated fans. It offered something for nothing: free games—interrupted, of course, by the most creative car-and-beer sales-jobs money could buy and often restricted when a game wasn't sold out, but still frequent and free—to those who didn't care to belly into the ballpark. If attendance suffered, the loss was typically more than offset by owners' share of TV advertising rights. But along that happy highway, Comcast and its peers have set up a lucrative tollgate, crowding out would-be satellite, broadcast, and Internet rivals and collecting a fat stream of dollars from viewers as well as advertisers.

Pro sports, with their physical heroism, taxpayer subsidies, and protected monopolies, have attracted their share of hard-charging moguls, who buy a team to enhance their status and proceed to project themselves as aggressive, egomaniacal bullies, grabbing credit for their athletes' triumphs and lambasting coaches or even players when things go poorly. But the Robertses are pleased to leave such credit and blame to others. Just as Ralph Roberts, who built the nation's biggest pay TV company, and Steve Burke, who runs it, don't much watch TV, Brian Roberts, who dominates pro sports programming in three of the nation's five biggest markets, isn't much of a pro sports fan. To Comcast's controlling owners, controlling Philadelphia sports is a means to stoke their real passion: growing their business.

Bullies on Broad Street

Ralph Roberts resisted early invitations to invest in sports. Comcast's rapid moves to dominate hometown sports in Philadelphia, Washington, Chicago, and other cities date to the mid-1990s, and to Brian's increasing involvement and self-confidence in his vision for the job he grew into. Hockey games and boxing matches were among the first programs televised by cable pioneers in New York and other big markets. But Philadelphia pro sports were slow to get wired. For one thing, the city took its time choosing which cable monopolies would control home viewing of the Phillies, Eagles, Sixers, and Flyers. Elsewhere in the region, the cable business was fragmented; Comcast only became the dominant provider in the late 1990s. Also, programmers lacked certain incentives: In the 1970s, most Philadelphia teams were a national embarrassment. Only after the brawling Philadelphia Flyers bashed their way to improbable back-to-back National Hockey League championships, in 1974 and again in 1975, did Flyers owner Ed Snider step forward and take the initiative to set up the Prism network, which wired games to fans in those advanced suburbs that already had cable.

That was before ESPN and round-the-clock sports shows. Prism, like other sports pioneers, filled its dead hours with Hollywood movies; it competed with HBO and other movie channels, and with enough capital might have served as the base of a programming empire. Snider, whose day business was operating the nation's biggest stadium management company, SMG, had gained control of the city's indoor arena, the Spectrum, at the expense of his ex-mentor, former Eagles owner Jerry Wolman; he added the woeful Spectrum-based 76ers of the National Basketball Association, as well as the city's oldest and losingest sports franchise, the baseball Phillies, then in nearby Veterans Stadium, to his upstart network. Prism was promising, but ad revenues were still a fraction of their potential; many games in those days were also carried on free TV, since advertisers still had no other way to reach the city dwellers who remained the core of the local fan base.[5]

There were efforts to make Prism something larger. Gerry Lenfest, the former Annenberg lieutenant who hoped to concentrate Philadelphia-area cable in his own hands, called on Ralph Roberts in a vain effort to get his backing for Prism. But Philadelphia wouldn't buy, and

Snider, like many ambitious Philadelphians before him, finally turned to New York. Cable pioneer Charles Dolan had begun wiring Manhattan twenty years before Comcast got into Philadelphia; he later wired the New York Rangers live from Madison Square Garden twenty years before Comcast made its own move into sports programming. From its power base, Cablevision was an investor in systems all over the country, including Philadelphia, and tinkered in many TV-related businesses. In 1983, Snider sold Prism to Cablevision. He kept the arena, and the Flyers.

So You Want to Own a Pro Sports Team

For a few frantic years, Pat Croce got to live the middle-aged American's dream: He ran a pro sports team, the Philadelphia 76ers of the National Basketball Association, and helped make it a contender. He was forced out in the end, when his ambition outgrew his place in Comcast's corporate pyramid. But he remains grateful to his ex-employers. "It's a cult, but a good cult," Croce said. "They control the world," he added, speaking of Ralph and Brian Roberts. "Luckily," he hastened to say, "they're good people."[6]

In 1996, Croce had written a check he couldn't quite cash. He had convinced diet-food king Harold Katz to sell him the struggling Sixers at a price he just couldn't swing on his own. Now he was beating the bushes, trying to find a deep-pocketed Philadelphian to back his bet. Like many who travel the well-beaten path from blue-collar parish to leafy Main Line suburb, Croce was a wiry, energetic, over-the-top Philly kid, who went out of his way to emphasize his genuine Philly street credibility. Croce studied as a sports therapist, parlaying his skill and promotional acumen into a chain of physical-therapy centers that catered to the aches and pains of dedicated athletes as well as affluent, aging baby boomers. He met the Robertses through real estate impresario Ron Rubin—or, more precisely, through Rubin's wife, Marsha, who was anxiously trying to avoid the orthopedist's knife. "I said, 'You don't need surgery,' and I got her new Reeboks. And she was so grateful! She said, 'You gotta meet my Ronnie!'"

At the time, Brian Roberts was facing a different problem: Cablevision's Prism, which he had to pay to keep hockey, baseball, and basketball on Comcast. While ESPN was offering round-the-clock shows, Prism seemed content to milk its monopoly on games. Philly's team owners began grousing that they could boost their revenues with more aggressive promotional programming, before-and-after shows, commentaries, sports-related video. Brian Roberts agreed. But who would make it happen?

Knowing that Croce needed money for his ambitions, while the Robertses needed talent for theirs, Rubin took on the familiar role of business matchmaker, bringing Croce to meet Ralph and Brian at the Hospital of the University of Pennsylvania's 1996 Christmas party. "Ralph comes in. The patriarch of Philadelphia business. And it turns out Ralph has knee problems," Croce exulted. "I said, "Take your pants off." They went into a bathroom. "Brian comes in. His dad is on the bathroom floor, and I'm working on his knee."

Comcast's Prism contracts were starting to come up for renewal when Croce approached, hat in hand, looking for a team to run. The Robertses assigned attorney and dealmaker Arthur Block to look at the deal.[7] Block met with Katz and his lawyers, hammered out the details, then proposed a multisided deal to Flyers owner Ed Snider, who shared the Spectrum with the Sixers' Harold Katz: Let Comcast take majority control of both indoor teams. Snider would continue to run the Flyers; he would still own a third of the team, and gain a chunk of the Sixers. The energetic Croce would step out front and run the Sixers, owning 2 percent and managing their high-priced, high-profile young players and their big, brittle egos. Croce would be the Sixers' public face: He would get the praise, and the blame; many fans would assume he, in fact, owned the team.

"We could have bought one hundred percent of the Flyers and Sixers, but I didn't want to be the owner making the tough calls," Roberts later said.[8] As it was, Comcast's Spectacor sports affiliate would account for "less than one percent of our business, but [it] takes up about forty percent of our time."

The Robertses, in their usual executive management style, gave Croce plenty of room. As Brian put it at the time, "I leave the high profile decisions to Ed and Pat."

"I never saw them unless I called them," Croce confirmed. "They let me do anything and everything with the Sixers. I told 'em we needed a big gun like Pat Riley in Miami—they hired Larry Brown for a bundle." That made them unusual among the flamboyantly selfish tribe of pro sports owners. "Their self-confidence, their self-esteem, is so strong they don't need fame," Croce said.

Playing in Comcast's new arena, the Sixers were a lively crew that came from nowhere to come as close as anyone did to overthrowing the LA Lakers' dynasty, which had dominated the league. Croce bridged the gap between Comcast's old-rich, quiet, buttoned-down Jewish owners and its newly rich, turbulent African American talent; in 2000, the team led by explosive Allen Iverson made it to the finals, against the LA Lakers, and Croce was briefly a household name.

The Real Owner

Even though Brian Roberts's company controlled the cameras, it's doubtful a TV color commentator covering a momentary lull in the increasingly frantic scene at Sixers games would have rolled on the happy owner in his special box or midcourt seat. That's because, with few exceptions, he wasn't there. "We got courtside seats for Brian, and he gave them away," Croce marveled. When the Sixers won the Eastern Conference title, Croce had to push Roberts onto Action News so "they could interview him—the real owner."

Though you can see the South Philly stadiums from the back of Comcast headquarters if you look down 15th Street on a clear day, the Robertses run Comcast–Spectacor as they did QVC: with clear goals but hands off, unless some strategic move with outside parties needs to be negotiated or finessed. At corporate headquarters, the financial, legal, and senior operating people can disagree, within bounds. Ralph doesn't raise his voice; Brian speaks faster and with greater intensity as the questions get tougher, and is not above ridicule. But field executives are expected to solve their own problems: "I never saw them," Croce said, "unless I called them." Is the low-key front a form of risk aversion? "They like being under the radar screen. They don't like being up front in bad times," Croce added. When possible, they avoid ultimatums and open "hardball," preferring to negotiate steadily and relentlessly toward a solution.

But when it came time to make his next big move, Croce didn't copy Roberts ways; he tried the playbook of his immediate boss, Ed Snider. Snider had become a power in Philadelphia sports a generation earlier by moving out from under the shadow of his own boss, the well-liked but financially embarrassed Eagles and Spectrum owner, Jerry Wolman. Croce wanted more than the Sixers presidency; he wanted to run Comcast–Spectacor. Snider, for his part, noted Croce's popularity—or at least the public impression that Croce owned the Sixers—and took to stressing the "promotional" nature of Croce's service.

It all came to a head in the summer of 2001. Croce appealed openly, through the city's lively sports media, for the CEO job at the sports-and-stadium division. Snider refused, and the Robertses declined to intervene. The numbers favored Snider: He owned a third of Comcast–Spectacor; Croce, around 2 percent. Croce resigned.

On a recent visit to his office on Philadelphia's Main Line, Croce touted his new careers as a motivational speaker and sports promoter. He has since gained his own reality TV show, *Moving In,* and is setting up a pirate museum in Key West. Snider's own prospects beyond Philadelphia are vast: Having sold the SMG stadium-management business, the nation's largest, which he ran for partners Aramark and Hyatt in 1997, he has reentered the business by setting up a new management company to run minor-league, college, and small-city stadiums, a business that could become far greater if Comcast ever buys Disney and grants Snider a role in that company's vast facilities-management group.

Comcast's late 1990s expansion into Maryland enabled it to set up a similar arrangement in the large and growing Washington, D.C., metro market, under Comcast SportsNet President Jack Williams, a hard-working, humorous cable veteran who started his cable career at Irving Kahn's TelePrompTer, ran Philadelphia's Prism in the 1980s, and was considered a natural to run the twenty-four-hour operation when Comcast went in with Snider.[9] The company's next pro sports targets, Chicago and Los Angeles, will give it a dominant position in four of the top five media markets.

Come and Play

If all it took to attract the Olympics was a neighborhood full of sports venues and parking lots, South Philadelphia would bid fair. At the city's lower tip, below the rowhouse blocks, where Interstate 95 bends between the green bridge to Jersey near the spot—by airport, refinery, and shipyard—where the Schuylkill dumps into the tidal Delaware, sit four arenas: the open-air Lincoln Financial Field and Citizens Bank Park, and the indoor, Comcast-owned Wachovia Center and Spectrum. The city's long-suffering taxpayers mostly built the new outdoor venues, though the owners of the Eagles and Phillies were allowed to sell the naming rights. The Robertses, like most urban builders, took advantage of tax breaks and site acquisition help, but they raised their own building costs and even arranged for Philadelphia's biggest bank, CoreStates, to kick in $40 million (over a twenty-nine-year period) for naming rights for the new and larger indoor arena, the Center, next door to the aging Spectrum. Bank mergers later made this venue the First Union Center—which Philadelphians quickly shortened to "F-You"—and, finally, the Wachovia Center, or Wack, for short."

But Comcast also had its own signs featured prominently on the building, promoting itself and various visitors, from the X Games to the 2000 Republican National Convention, when Comcast was able to use the house it built to promote itself to a first-class national audience.

14
MICROSOFT
BUYS IN

"As Monopolistic as You Can"

Intense, tailored Brian Roberts well fits his role as the new public face of cable TV. He has a small, warm smile; a rapid, level way of talking about his business in a dispassionate, finance-techy, half-slang storm of words; profound self-confidence, kept manfully in check, balanced by his competitive jitter that someone, somewhere, might be gaining on him; and apparent reluctance to speak ill of others by name in public. To appreciate all this, consider the funny, belligerent, grasping gladiators who filled this role before him: first, franchising pioneer Irving Kahn, who went to prison for bribery, then built a second cable empire to the cheers of peers who felt he'd done nothing wrong; next, satellite booster Ted Turner, dubbed "Captain Outrageous" on *Time* magazine's cover, who baited Christians, insulted politicians, and left his wives to date movie stars; and, especially, deep-thinking, cheerfully profane John Malone—"Darth Vader" to his fellows. He once cut off television to Vail, Colorado, to force city council to his terms, seemed never to pay taxes nor to improve a wretched mile of his rundown TCI cable network if he could possibly avoid it, and flippantly told *Wired* magazine that shooting the

FCC chairman would improve the business climate. He also called Senator Al Gore a skunk, giving the senator and his publicity-seeking colleagues such a great political target that, when the Democrats gained power at the beginning of the 1990s, they passed a rule forcing cuts to basic cable rates (which helped provoke a boom of cable companies buying noncable businesses).[1]

The Yale-educated son of a Pennsylvania engineer, Malone threw away careers as a cutting-edge scientist at Bell Labs, a hotshot management consultant, and the head of Milton Shapp's old cable-equipment maker Jerrold Electronics to run Denver-based TCI, which was for a time the nation's biggest cable purveyor—and the stingiest, his many critics argued. Reminded by Gore how a Malone company had once been fined $25 million for threatening to "destroy" his rivals in a contentious Missouri franchise fight, Malone replied with delicious candor: "In our society," he testified, the need to make money for your shareholders "leads you in the direction of trying to become as monopolistic as you can."[2]

John Malone helped Ralph Roberts from an early date, and took Brian in hand to show him the ropes. They served together on the board of Turner Communications in the late 1980s, when Brian was still in his late twenties. At his first big opportunity, young Roberts repaid Malone's trust with what his would-be mentor regarded as a shocking treachery.

Malone had worked cordially enough with Brian's dad. In 1986, Comcast and TCI had split parts of the former Westinghouse cable network. That same year Ralph Roberts invited TCI to back his investment in QVC, spreading the risk; six years later, Malone and Roberts backed then-QVC chief Barry Diller in his initial attempt to seize Paramount Pictures, and Malone chose Roberts over Diller in their ensuing split. (This provoked rival bidder Viacom to file a lawsuit in which it accused Malone of building his career with "bully-boy tactics and strong-arming competitors, suppliers, and customers" in order to grasp "monopoly power" over cable programming, satellite transmission, and home service: "At every stage of the process, the consumer has paid and will continue to pay a monopoly tax to John Malone," Viacom charged.[3]) TCI and Comcast joined forces again to support General Instruments' mid-1990s turn toward digital TV equipment—to their immense profit when that company was sold to Motorola at the peak of the market in 1998.[4] Malone also joined Roberts in the famous meeting with Bill Gates in 1997, when the software statesman agreed to invest $1 billion in Comcast.

Learn and Do

Along the way, Malone found time to help educate young Brian Roberts about the shape of media to come. On a trans-Pacific flight to a Japanese electronics show in 1996, Malone kept Brian enthralled with his vision of cable Internet, finance, technology, and taxes. "We talked about the future of the world, and it was fabulous," Roberts told Malone's biographer, Mark Robichaux, of the *Wall Street Journal*. "It was like going to business school."[5]

If Roberts was awed, he wasn't cowed. In the fall of 1997, the sons of Malone's late boss, Bob Magness, sued Malone for low-balling their TCI stock when he'd bought them out after Magness's death. As proof, the suit cited an unidentified bidder who had offered a better price. Baffled, Malone tried to identify the mystery buyer. Leo Hindery, Malone's second-in-command, told Robichaux that Brian Roberts approached him at an industry gathering at Morton's, the fancy chain steakhouse, in New York that October and acknowledged that Comcast had been the mystery bidder—presumably with the support of Microsoft's Bill Gates, who was apparently not content to let his cable influence stop with just Comcast, the number-three player in the business.

"Brian," Hindery begged, "how can you do this?"[6]

"Malone was livid," Robichaux reported. "Business was business, but Malone viewed this offense as an unforgivable breach of what was right": in Malone's words, the "underhanded way Roberts and Gates had gone about it. . . . They screwed me."

"Brian broke the code," added Hindery. "I will make that son of a bitch pay." Brian Roberts apologized, and so did Bill Gates—both to Malone, if not to Hindery. "John and I sat down, we had one long night in Sun Valley," where cable executives gather each fall, Roberts told *USA Today* four years later. "I said I feel bad, I'm sorry. And he said it's behind us."[7]

Malone later mocked Roberts for his "bended knee" subservience to Gates[8] even though the Microsoft–Comcast relationship has proved highly profitable to both companies and their peers. He also bedeviled Gates's cable TV plans, favoring Gates's competitors over Microsoft as the developer of digital-cable boxes. But whatever his personal feelings, Malone continued to do business with Comcast, again to the profit of both companies. Roberts and Malone might argue; yet they remained members of the cable fraternity. To be sure, they have sued each other, notably over the cost to Comcast of Malone's Starz movie channel; and they've done business, in particular, the summer 2003 deal in which

Comcast sold control of QVC to Malone's Liberty Media, giving Comcast the capital to show its biggest profit ever for that year, and to go on and make an offer for Walt Disney Company the following winter.

What Competition?

Frat reunion, pep rally, trade show, nerd circus: No other industry in America still congregates, top to bottom, with quite the gusto of cable TV's operators, their dependent businesses, and the growing crowds of channels and gadget developers. "We are the only vertically-integrated industry left in America, and we like to get together," said Frank Drendel, chairman of Commscope, the nation's largest cable manufacturer.[9] In national and regional conclaves, the cable operators assemble, joined by all those beholden to them: suppliers, programmers, regulators, financiers, lobbyists, the trade press. These days-long gatherings at big hotels illuminate successful egos, establish pecking orders, argue competing ideas, promote deal making, and generally cement, or fray, the personal bonds that hold the industry together. "Except when we're actually fighting each other, we're all friends," said H. F. "Gerry" Lenfest, who switched career paths, from cable baron to philanthropist, when Ralph and Brian Roberts forced him to sell in 1999.[10]

"It's like down South," Drendel said: We may like to fight among ourselves, but "if anyone comes on our farm, we all get together and drive 'em out."[11] There used to be a Cable Club in Philadelphia, and regional cable shows around the country; those have faded amid consolidation, but that just makes the national gatherings that much more important.

Message from Bill

In 1996, when Drendel was head of manufacturing at General Instruments' cable division, Milton Shapp's old company, he found himself rushing excitedly in the company of GI sales chief Ed Breen (whose top clients included Comcast) through yet another national cable show to meet one of the richest men in the world, William H. Gates Jr., founder and chairman of Microsoft, at Gates's own request. Gates had a soothing message, as Drendel recalled it: "He said, 'Cable is going to beat telephony. You guys are absolutely the pipe to the digital future.'"[12]

Gates's own tendencies toward monopoly have kept some of the nation's finest legal minds employed at hundreds of dollars per hour for many years. In the second half of the 1990s Gates had come to see TCI,

Suzanne Fleisher Roberts, a flamboyant stage, radio, and TV personality, in 1952. Her family and its connections in Philadelphia's Jewish business community boosted husband Ralph Roberts's business ventures, including Comcast. Their enduring marriage produced five children, including Ralph's successor, Brian. Philadelphia Inquirer *News Research*

Ralph Joel Roberts, in 1980. At the time, Comcast was one of dozens of family-owned cable companies that looked like deal-bait for big corporations. Unwilling to depend just on cable, Roberts's company still dabbled in other businesses, such as canned music and recorded retail advertising. *Comcast Corporation, 1980 annual report*

Ralph Roberts, *left*, and Comcast's chief financial officer, Julian Brodsky, in 1982. A labor unionist and Democrat in his youth, Brodsky learned to love capitalism as an auditor for little family-run cable systems in upstate Pennsylvania; deeply impressed by Roberts, he joined Ralph's nascent company. Architect of Comcast's creative financing arrangements, Brodsky later made millions in Internet investments. *Comcast Corporation, 1982 annual report*

Brian Leon Roberts, in 1984. The only one of the five Roberts children to show much interest in working for his dad, he followed Ralph through the University of Pennsylvania's Wharton business school, then demanded a Comcast job. His keen interest in his dad's work encouraged Ralph to turn Comcast from a grab bag of businesses into an aggressive media giant for Brian to inherit. Philadelphia Inquirer *News Research*

Left to right: Comcast owner Ralph J. Roberts, in 1984, with two of his first and most important hires—Daniel Aaron, a newspaperman turned cable broker, manager, and lobbyist; and financial wizard Julian A. Brodsky—at the company's closed-circuit broadcast studio. Like other cable companies, Comcast first charged customers to transmit broadcast TV programs that it collected for free, then added proprietary programs (sexy movies and sports) and systems (scrambled and digital signals), which it could refuse to share with other media, so it could charge customers more. *Gerald S. Williams*, Philadelphia Inquirer

Left to right: Daniel Aaron and Ralph Roberts of Comcast thank smiling Philadelphia City Councilman Francis Rafferty for helping Comcast secure the city's most lucrative cable franchise, in 1984. To clinch the deal with the city's ruling Democrats, Comcast promised to move its headquarters from the suburbs to Center City and sell nonvoting shares to African Americans. *Gerald S. Williams*, Philadelphia Inquirer

Ralph Roberts kisses his wife Suzanne at the 1984 Philadelphia City Council hearing where Comcast was awarded a city franchise after eighteen years of false starts and persistent politicking. *Gerald S. Williams*, Philadelphia Inquirer

Ralph Roberts dons a ceremonial hardhat to commemorate Comcast's success at finally connecting Philadelphia, the last major city wired for cable TV, in 1986. *Ron Tarver*, Philadelphia Inquirer

Comcast's management had a strong family cast before Brian joined its ranks. Directors have included Ralph Roberts's brother-in-law, Robert Henry Fleisher, and a Fleisher son-in-law, Sheldon M. Bonovitz, who rose to the top at one of Philadelphia's big law firms; Ralph's brother Joseph ran Comcast's Storecast division before his early death. Ralph's son-in-law, Anthony Clifton, was briefly a vice president. Comcast's board of directors, as pictured in 1988, *left to right:* investment banker Joseph L. Castle; Sheldon M. Bonovitz; real estate investor Gustave G. Amsterdam; Comcast executives Bernard C. Watson, Brian L. Roberts, Ralph J. Roberts, Daniel Aaron, Julian A. Brodsky; and Pittsburgh cable investor Irving A. Wechsler. *Comcast Corporation, 1988 annual report*

Ralph, *seated*, and Brian Roberts pose for a portrait, in 1990. After hanging around the office as a schoolboy, sitting in on his father's deals, and compiling a mixed record in entry-level Comcast jobs, the boss's son held a string of increasingly senior management posts. Still in his twenties, Brian sharpened his negotiating skills with Wall Street investment bankers; he built ties to powerful politicians and key media executives for Comcast in Washington, D.C. *Rebecca Barger*, Philadelphia Inquirer

Left to right: Philadelphia Mayor (later Democratic National Committee chairman, Pennsylvania governor, and Comcast sports talk-show host) Edward G. Rendell, with a furry tennis mascot; Pennsylvania investor Warren "Pete" Musser, who sold Ralph Roberts his first cable TV system and the QVC home-shopping channel; Brian Roberts; and pro tennis player Vitas Gerulaitis, at a 1993 promotional event. Philadelphia Inquirer *News Research*

Left to right: Ed Snider, who traded control of the Philadelphia Flyers to Comcast for a job running Comcast's Philadelphia sports monopoly; Ralph Roberts; Harold Katz, former owner of the Philadelphia 76ers; Pat Croce, the physical therapy entrepreneur who talked his way into a job as the 76ers' popular and energetic front man for a time; and Brian Roberts, in1996. Ralph and Brian Roberts were atypical pro team owners, who set tough business goals but left the glory of victory and the agony of day-to-day operations to their handpicked managers. *George Miller*, Philadelphia Daily News

Ralph, *right*, and Brian Roberts signaled their ambition to dominate television and Internet media through their hostile 2001 offer to buy AT&T Broadband. In a stock market downturn, they outmaneuvered AT&T Chairman Michael Armstrong to acquire his company's lucrative cable systems and promising new technologies for a bargain price. *Tom Gralish*, Philadelphia Inquirer

Suzanne Fleisher Roberts, in a 2002 publicity photo. As an octogenarian, the founder's wife, and the boss's mom, she starred in a soft-focus, five-minute daily program, *Seeking Solutions with Suzanne*, which touched on topics like sex, art, and face-lifts, all for senior citizens. Philadelphia Inquirer

Brian Roberts, *center*, with his neighbor, the actor David Morse (*Hack*), from Philadelphia's wealthy Chestnut Hill, catches a rare 76ers game in 2003. Roberts and his dad owned less than 2 percent of Comcast's stock but exercised veto-proof voting control under special rules that drew protests from dissident shareholders; their hand-picked board paid the pair over $250 million from 1998 to 2003. *George Reynolds*, Philadelphia Daily News

Brian L. Roberts. His acquisition of AT&T's cable arm made him the most powerful man in television, giving Comcast weight with programmers, advertisers, viewers, and government officials. After his bold February 2004 offer to buy Walt Disney Company was turned down, he promoted Comcast as a one-stop home media company, combining Internet, phone, and video-on-demand services. *Comcast Corporation, 2004*

Ralph J. Roberts. Retired from day-to-day management after the AT&T deal, Ralph remained Brian's counselor and confidant, and had the pleasure of watching the son he had raised in the business expand his legacy through new takeovers and technologies. *Comcast Corporation, 2004*

Federal Communications Chairman Michael Powell, *right*, with giant remote control, helps promote Comcast's new high-definition TV services at a 2003 cable TV trade show in Chicago. With him is, *from left*, FCC Commissioner Kevin Martin, Comcast Communications President Steve Burke, Comcast founder Ralph J. Roberts, and Comcast Chief Executive Officer Brian L. Roberts. Though he blocked merger moves by rival satellite TV companies, Powell approved of Comcast's giant acquisitions and said he expected big companies like Comcast to invest in popular new video services; critics said he favored the industry at the expense of American consumers. *Oscar & Associates*

Comcast executive vice president David L. Cohen, *right*, with Comcast spokeswoman D'Arcy Foster Rudnay and public relations adviser Adam Miller of the Abernathy McGregor Group, monitors the 2004 annual shareholders' meeting of Walt Disney Company from Comcast headquarters in Philadelphia, a few blocks from the Disney meeting. Disney shareholders voted narrowly to retain Michael Eisner as chief executive of Disney; Eisner and his supporters on the Disney board also fought off Comcast's attempt to buy their company that year. *William F. Steinmetz*, Philadelphia Inquirer

Former U.S. Sen. George J. Mitchell, the key outsider on Walt Disney Company's board, addresses fractious shareholders at Disney's 2004 annual meeting in Philadelphia. Three years after Mitchell gave Comcast's Brian L. Roberts a glowing introduction for Roberts's speech to political, financial, and business leaders at the Washington Economic Club, Roberts was hoping Mitchell would buck Disney chief executive Michael Eisner and support Comcast's attempt to take over his company. Instead, Mitchell and other directors rallied to Eisner and rejected Comcast. *Eric Mencher*, Philadelphia Inquirer

TOP 10 REASONS COMCAST IS GREAT

10. You can watch the 2004 Summer Olympics in HD.

9. You can surf the Web up to 50 times faster than dial-up and 3 times faster than 768K DSL.

8. You get a clear, dependable picture — even when Mother Nature dishes it out.

7. You can make a turkey sandwich in the middle of the game and not miss a minute of the action.

6. Your kids can watch what you want them to watch and nothing more.

5. You can surf full-throttle with the power of 100% Pure Broadband.™

4. You can tell Tony Soprano when to show up in your living room. And he'll be there.

3. You can record a whole season of Everybody Loves Raymond with just a few clicks of the remote.

2. You'll never have to pay another late fee at the video store again.

1. More than 21.5 million customers can't be wrong.

Not all services available in all markets.

Top Ten Reasons Comcast Is Great. Comcast has always had other businesses in addition to cable TV. As new basic cable subscriber growth slowed in recent years, the company has promoted new services like high-speed Internet, phone, and video-on-demand. *Comcast Corporation, 2004*

Comcast used this happy-looking (but fatherless) family in a campaign promoting its media services. Although critics like George Gilder (*Telecosm*, Free Press, 2002) have called TV "the supreme time waster," "sordid," and "sleazy," it remains universally popular. "People want to watch television, and more is better," Ralph Roberts told the *Philadelphia Inquirer* in 1994. "What's the first thing they do when they come home at night—kiss their wife, maybe? And then run over to the television set." But asked if he could name a favorite show, Roberts said, "Not really. I'm the worst television viewer there is."

Ten years later, Brian Roberts celebrated the convergence of Comcast's fast-growing Internet service with its cable TV and video-on-demand. "The way people interact with television is absolutely going to change," he said in announcing the Disney deal. "We are trying to give consumers choice. That's who's going to win." But Roberts's distant cousin-by-marriage, Internet scholar Cass R. Sunstein of the University of Chicago (*Republic.com*, Princeton University Press, 2002), has warned that viewer-directed video and Internet risk creating ghettoes of the mind: "Tens of millions of people are mainly listening to louder echoes of their own voices." *Comcast Corporation, 2004*

"Monopoly money," featuring the faces of Comcast executives and listing concerns about corporate governance, programming, prices, customer service, and political deal making. The Philadelphia Community Access Coalition (www.phillyaccess.org) first distributed these at a Monopoly-themed rally targeting Comcast in 2001. The coalition organizes city residents to push for public-access cable television as

10 Reasons to fight a Comcast MONOPOLY

They:
1. Charge Philly the highest cable rates in the country
2. Deny our right to Public Access TV
3. Want to be the biggest broadband provider in the world
4. Oppose Open Access to the Internet
5. Monopolize local sports
6. Drive out competition
7. Make huge campaign contributions to politicians
8. Close community television studios nationwide
9. Lobby against the public interest in Washington DC
10. Provide lousy customer service

www.phillyaccess.org • 215-563-1090

Comcast Lobbies for MONOPOLY

Comcast Media Merger Lobby Team connected to White House, Defense, Congress

- **Lorine D. Card**, Sister-in-law of White House chief of staff Andrew Card
- **Victoria Clarke**, Former Asst. Secretary of Defense for Public Affairs under Donald Rumsfeld
- **David L. Cohen**, Former aide to (now) Pennsylvania Governor Ed Rendell
- **Kerry Knott**, Former chief of staff to former House Majority Leader Dick Armey and lobbyist for Microsoft
- **Melissa Maxfield**, Former head of Sen. Tom Daschle's political action committee
- **Alfred Mottur**, former senior telecommunications advisor to Sen. Ernest Hollings, Commerce Committee member
- **Jessica Wallace**, Former senior advisor to Rep. Billy Tauzin while he was chair of House Energy and Commerce Committee

According to Comcast's lead political operative, **David L. Cohen**, this lobbying team gives the company "a lot of balance. Republican and Democrat, House and Senate, people who have relationships throughout Washington."

Sources: Multichannel News, Sept. 29, 2003; Wall Street Journal, Feb. 13, 2004.

www.mediatank.org • www.phillyaccess.org • www.democraticmedia.org

---

Why are you paying the highest cable rates in the nation?

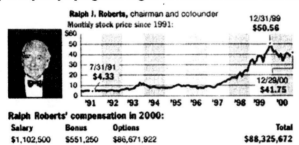

Ralph J. Roberts, chairman and colounder
Monthly stock price since 1991:

12/31/99
$50.56

7/31/91
$4.33

12/29/00
$41.75

Ralph Roberts' compensation in 2000:

Salary	Bonus	Options	Total
$1,102,500	$551,250	$86,671,922	$88,325,672

Comcast executives get rich playing MONOPOLY with your media!

---

provided for in the city's own cable TV ordinance. Media Tank (www.mediatank.org), a Philadelphia-based media education and advocacy group, later adapted the funny money for use in other media campaigns. *Media Tank / Philadelphia Community Access Coalitio*n

After national surveys found a public backlash against high cable rates and poor service, competing EchoStar Communications and its DISH satellite TV network invited cable viewers to switch to satellite. In early 2004, EchoStar stationed giant pig-shaped balloons at Comcast offices in Pennsylvania, Colorado, and other states, and ran ads urging viewers, "Don't feed the cable pig!" Comcast and other cable companies fought back, ridiculing satellite's vulnerability to weather. To protect their old monopoly status, the cable giants also promoted high-speed Internet, video-on-demand, cable-phone, exclusive sports, and other services they hoped satellite couldn't match. Comcast even suspended Brian Roberts's cherished high-fee policy and reduced some of its rates, for a time, to attract new users. *EchoStar Communications Corporation*

Comcast, and their operating peers as key players in yet another market he hoped to conquer; but before he went to work on the operators who held the key to his plans, he had wanted to make sure the technology could handle what he had in mind. If Gates liked cable, the feeling was mutual, at least in some corners. In 1997, John Malone "organized a cadre of cable cowboys for a technology tour" to a number of West Coast technology companies. In Mark Robichaux's telling, Comcast owed Malone's investment to "a behind-the-scenes courtship that was initiated largely by John Malone." If so, Malone's matchmaking, for once, benefited others more than himself.[13]

Under the Gun

The cable operators needed some excitement, if only to perk up skeptical investors. They were "really depressed," Gates later told an audience at a New York gathering in 1998.[14] Cable stocks were plunging in the middle of the Internet rally. Compared to AOL and other Internet utilities, wireless, satellite, and even phone companies, cable was starting to look old and stale. And, as Gates put it, "They'd gotten themselves under the gun in a lot of ways": Regulators were threatening to cut rates for the third time in a decade. Rupert Murdoch, who with his newspapers and television stations held himself above the cable gang, was threatening to back satellite TV and force a price war—the worst thing a self-respecting cable operator could imagine. Cable had invested heavily in interactive TV, cell telephony, and other edgy technology, but had little to show for it. Even General Instruments had proven slow to deliver on the promised digital TV equipment Ralph Roberts had called for more than a decade earlier.

Gates saw the many causes of cable's frustration. But he was even more frustrated at cable's rivals, the nation's entrenched phone companies, who ought to have been leading the rush to digital America. They had plenty of money and plenty of lobbying clout, but they were weighed down with too much bureaucracy, too much old copper wire; not enough new fiber-optic connections existed, and apparently not enough energy or drive to wire America to high-speed Internet as fast as Gates thought the job should get done. He turned to cable, with its tradition of risk and brash aggression, as an alternative route into the nation's homes and businesses.

Gates invited cable-industry leaders to Seattle for a special summit in June 1997. His message was simple: Microsoft could write and license software for the set-top boxes made by General Instruments (or its few competitors, for that matter) to deliver digital programming "that your customers will find exciting."[15] Cable had to keep upgrading its network, get out there and promote the service, and make everyone who had a piece of the action rich. The cable barons seemed reluctant to take Gates's vision at face value: "They kept saying, 'Well you at Microsoft are so clever, do you really really think the cable industry has a great future?' And we said, 'Yes.'"

But who would pay? Maybe cable had the will to rebuild, which Gates found lacking in the phone companies. But the operators, heavily in debt from the costly takeovers that most of the companies had used to gain their current clout, said they lacked the cash for further investment.

10 Percent

What to do? It was June; the Pacific salmon run had begun a few weeks earlier. The robust native fish was Seattle's symbol of its vaguely natural- ist identity, its vigorously healthy lifestyle, but also of the cheerful preda- tion of its biggest businesses: aerospace and software. Naturally, Gates brought his visitors together for a fancy fish dinner at a club high above Seattle. Brian Roberts and his old mentor John Malone of TCI were among the cable leaders in attendance. Roberts and Gates knew each other only slightly, so Brian's boldness—and Gates's response—was all the more remarkable.

According to Roberts, he told Gates, "You know, what we could re- ally use is for you to buy ten percent of every company in this room."

According to Gates, Roberts said, "'Come on, are you really serious about that?' And I said, 'Look, if I need to take some of the cash we gen- erate and invest it in a passive way, and wanted to prove we believe in cable, I will do that.'"[16]

Roberts reconstructed the dialogue for a Penn alumni magazine:
"Gates hesitated, then asked, 'What do you think that would cost?'
"'Oh, I don't know, maybe $5 billion.'
"'You know, I have $10 billion in cash. I could do it.'

At that point, Roberts said he "swallowed hard." Someone else de- railed the conversation by asking Gates about his pending Amazon vaca- tion. But later Gates pressed on: "'Would there be a regulatory prob- lem?'"[17]

Later that week, Roberts said, Gates followed up, calling all the way from the Amazon, via Microsoft's chief financial officer, Gregory Maffei. Three weeks later, Gates said, Roberts joined him for breakfast with an investment banker, "and by the end of the day I'd agreed to make a passive investment in Comcast," buying $1 billion in stock to help in "updating their infrastructure for high-speed connections." The press made much of Gates's agreement to purchase nonvoting stock; shares rose sharply, and cable stock was on its way to yet another boom. But the voting status wasn't so unusual; this was Comcast, where Ralph Roberts had long ago ensured that most Comcast shareholders—except him—had little or no say.

Gates was careful not to oversell his commitment to cable. Like Comcast, which, for all its historically close business and financial relationships with General Instruments, preferred to buy equipment from multiple suppliers, Microsoft liked to play competing industries off against each other, enabling itself to benefit from competition while it offered as little as possible to its own customers. Microsoft "will work with anybody," Gates concluded. "But we think the cable companies have a very important role to play."[18]

MARKET POWER

Your Money, Our Choice

Cable operators don't like to be called monopolists. Nobody's forcing you to spend $600 or $900 or $1,200 a year on their services. You can choose instead to watch rabbit-ears TV, or satellite-dish TV, or Internet streaming video, they say. Or you can wait for your phone company or your electric company or your wireless service to add video. Someday.

Sure, you can't get a full menu of big-league sports programs, in some of the biggest metro markets, without buying the local cable service. But that's your choice.

And choice, Brian Roberts says, is what Comcast is all about.

That's what he tells the government. Buying AT&T would let Comcast "expand the range of information and entertainment services available to many consumers and provide those choices more quickly," he said in a statement to the U.S. Senate's antimonopoly subcommittee in April 2002.[1] He was introduced by Pennsylvania's Sen. Arlen Specter, a longtime Roberts family friend and major beneficiary of cable industry cash.

That's what he tells investors. "We believe that the way people interact with their televisions is absolutely going to change in the near future. It's a competitive business, and we're trying to give consumers a choice. That's who's going to win."[2]

It's what he tells consumers, in junk-mail and phone-spam solicitations for add-on Internet and video services at low introductory rates.

Roberts believes the very existence of the Internet has validated him: "Cable was about creating more choice and more options for television. In those days, as we all remember, we had three or four choices. And when we first went around and tried to sell 20 channels of television, people said, 'Why do I need 20 channels? I can only watch one.' Then came 50 channels and a hundred channels. And I think the Internet today is the ultimate affirmation that American consumers want choice. There are millions of Web sites, and everybody wants to do something unique to their interest."[3]

You don't quite buy it? You don't see why you shouldn't be spared the yearly price increases at double the rate of inflation, or why you should be forced to buy channels you don't want to get the ones you do? Maybe the message isn't really meant for you, but for someone more important. Like former U.S. Sen. George Mitchell, for example. Mitchell believes Comcast stands for choice. Or at least he did in 2001, when he invited Roberts to address a roomful of power lawyers, lobbyists, and business executives at the Economic Club of Washington, which he heads, and told the crowd by way of introduction that Brian Roberts, like his father Ralph, understands that "first, people love television, and, second, cable is about creating more choices, and American consumers like choices."[4]

Four years later, as head of Walt Disney Company's outside directors, Mitchell would choose to reject Roberts and his company, despite all the nice things he'd said about them.

"Widespread Dissatisfaction"

The handful of towns blessed with real cable choices enjoy an average 15 percent break in their cable TV prices, the General Accounting Office announced in October 2003. In places where satellite competes with TV, by contrast, the cable price is just 5 percent lower than average. The GAO's conclusion: Cable and satellite don't quite compete.[5]

Two years before, the Federal Communications Commission traced the impact of cable competition, in the 2 percent of the country where it existed, in much greater detail and found that competing cable systems not only charge less, they also offer better channel selection.[6] Choice was good, the FCC found, in Alabama, where a public electric company "began construction of a new, municipally owned cable television system in Scottsboro because of widespread dissatisfaction" with the local

Cablevision affiliate. Choice was good in Georgia, where the local Du-luth cable monopoly "upgraded its system and added 19 new channels" in its "response to aggressive competition" when BellSouth's new cable arm moved into town.

And choice was especially good in parts of metro Boston, New York, Washington, D.C., and even a few Philadelphia-area outposts. There, RCN Corporation had buried hundreds of millions of dollars of venture capital in a state-of-the-art fiber cable system designed to make Comcast and the rest of the old cable cartel obsolete—or at least to make them sweat—to the benefit of the long-suffering cable consumer. Backed by investors like billionaire Paul Allen, Bill Gates's old partner at Microsoft, RCN offered what Big Cable had solemnly promised and utterly failed to deliver during the deregulation battles of the 1980s and early 1990s: a competing set of wires to give consumers a choice for their cable—and telephone and Internet—services. RCN threatened to break cable's most lucrative monopolies. It certainly sliced cable profit margins, as the FCC reported. In Massachusetts, in Virginia, even in a small slice of suburban Philadelphia's Delaware County, RCN's presence meant lower cable fees. RCN even had the audacity to come to Philadelphia and offer to wire the city's Northeast, where Comcast had gained its first big-city franchise just fifteen years before, absolutely free.

RCN was run by techies armed with modern fiber that could carry great volumes of data. If quality and economics alone determined who got to play, RCN might well have been able to offer Philadelphians the kind of productive price war that benefits everyone except entrenched monopolies. Yet RCN executives lacked the political skills the old cable companies had sharpened in the franchise wars, the rent-a-citizen days. And in Philadelphia, they'd be fighting Comcast, run by people who had shown the tenacity to wait twenty years for their first big urban fran-chise, outlasting mayors and council majorities and bidding wars to gain their prize.

Comcast had used its resources strategically, doling out modest grants to city projects like the Kimmel Center for the Performing Arts (though Verizon Corporation, for example, gave lots more), and other grants to key Democratic politicians, who had moved squarely into its corner. Comcast had given more than money: Hadn't the company stayed put, even added workers, through the whole grim 1970s, '80s, and '90s, when the city's banks, insurers, manufacturers, and railroads were closing down or leaving town? With its hard-won clout, Comcast wasn't about to give away its monopoly to someone who couldn't offer anything more than old-fashioned, free-market, consumer-friendly competition.

Philadelphia would see to that.

Get Outta Town

RCN proposed "to spend $250 million and create hundreds of jobs. Their project promised residents a choice for cable TV, telephone and Internet services. And RCN wanted no subsidies." They'd already been welcomed with open arms in Boston and Washington. "What wasn't to like?" asked the *Philadelphia Inquirer*.[7]

"Good God!" shot back then Mayor Ed Rendell. "We have to tear up the streets so you can come in here and compete against one of our best corporate citizens?"

"Pirates," City Councilman James Kenney called the upstart.

"They had this attitude," City Councilman Michael Nutter complained about RCN. Later, Nutter told the *Los Angeles Times* what he thought of Comcast: "It's hard to resist a parochial cheer at seeing a home-grown corporation for once be the predator, not the prey."[8]

There were dissenters. The Roberts family are "so powerful and well-connected, they could stop another multimillion-dollar business from creating new jobs in the city," objected veteran activist Lance Haver, who later became the city's consumer advocate under Rendell's prickly successor, Mayor John Street. But, in Philadelphia, Brian Roberts "is like a god," said Democratic political consultant Larry Ceisler. "They're star-struck."

A god, but with lobbyists instead of angels: Comcast hired Rendell's right-hand man, David L. Cohen; Nick Maiale, the South Philadelphia ward leader who heads the Pennsylvania state pension plan, a Comcast investor; and Democratic lobbyist Obra Kernodle, who said she was "shocked" that RCN had not "worked" city council to plead its case. Comcast assembled a report deriding RCN's effort to "digitally divide our city," and warned about the "high-tech, profiteering interloper."

The delay in approving the RCN deal stretched to two years, then three, until RCN ran out of cash and gave up. "Most cities would be thrilled to see the arrival of a new cable company that promised to invest millions in the community and bring sorely needed competition to the local monopoly," wrote the *Los Angeles Times*.

"In Philadelphia, city leaders ran such a would-be rival out of town."

That's how tough Comcast's friends were on Comcast's enemies: Philadelphia was a closed city.

Comcast allies also managed to get an RCN franchise effectively withdrawn in Prince George's County, Maryland, after Comcast hired a disbarred lawyer and longtime statehouse lobbyist, Bruce Bereano, to convince his former boss, the county executive, to reconsider. "I thought it was a done deal," a member of the county council majority, who supported RCN, said afterward. "We now have one provider, and it's the consumer who suffers."[9]

Doesn't Like to Share

Philadelphia cable pioneer turned billionaire philanthropist Harold Fitzgerald "Gerry" Lenfest has known Ralph Roberts so long he remembers the older man's doubts about the future of cable TV.[10] Lenfest came to dominate Philadelphia cable. Why, then, is Comcast the last heavy left standing? "Ralph Roberts always had this vision—I didn't share it" of going national, one careful deal at a time, Lenfest said. "He stuck with it. [Comcast] evolved into a good business because they're not wheelers and dealers. They took risks, but not outside cable."

Ralph didn't get that way by sharing. "We talked about merging several times. It didn't happen," Lenfest said. "One, I wanted to maintain my independence. Because Comcast was also in Philadelphia, if we [merged,] all our senior management would be terminated. Two, I have two sons, and I wanted to keep up the option in case one wanted to come into the business. But Ralph made it clear he had his eye on Brian. There was no room for sharing authority" with Lenfest, let alone his sons.

Don't get him wrong: "Personally I've always gotten along with Ralph. Good friends, good company. I have the greatest respect for Ralph, Brian, and Julian," Lenfest said. "[Steve] Burke [who joined Comcast from Disney in 1998] is a great operator. Brian is smart enough to delegate a lot."

Lenfest, by contrast, sought partners. In 1982, Lenfest had invited TCI chief John Malone to speak before a Philadelphia cable audience. Malone ended up buying 20 percent of Lenfest, a stake that rose to 50 percent under TCI's eventual buyer, AT&T. That support from a fellow member of the cable gang gave Lenfest an edge in his Philadelphia-area expansion.

But none of these men were in business for friendship or loyalty. As big companies bought small ones across the U.S., it was clear to Lenfest by the early 1990s that "the only way to survive was to grow." He kept absorbing systems, hoping to reach critical mass by piecemeal acquisitions. Much of the motivation was personal. By now, "my sons were no longer interested. But I didn't want to hurt our managers" by selling to someone who'd probably fire them. He and some of the other owners were getting old: "These guys wanted to cash out. They'd all have stock, which is useful for estate planning"—it's easier to avoid taxes on stock than cash.

Malone had been a hands-off investor. "Nobody from TCI ever set foot in our premises. The board never met in twelve years." The board consisted of Lenfest, his wife Marguerite, and Malone. But when Malone sold his national cable system to AT&T, at the top of the market in 1998, Lenfest's days were numbered. He developed a bold, half-desparate gambit, a final plan to stay independent: "In 1999, I wanted to buy all the private cable operators" in the region "and set up a public company." There would be the Walson family's Service Electric up in the old coal regions; Susquehanna Broadcasting in York; Harron in Chester County. "It seemed like a great idea. But I needed AT&T's approval. And they said no."

Leo Hindery, Malone's top lieutenant, now spoke for the corporation and not just for Malone, its biggest shareholder. "Hindery wanted me to sell" to AT&T, Lenfest recalled. "I told him I'd sell only with the stipulation, you cannot sell to Comcast within five years. If you do, you will owe me an additional $300 million. He told me, 'Put it in, I hate Comcast!'" Lenfest figured that was out of loyalty to Hindery's boss, Malone, whose company had been the target of a stealth takeover attempt by Brian Roberts and Microsoft just two years before.

"I was so convinced [Hindery] would never sell [to Comcast] that I forgot to put the provision in," Lenfest said.

But AT&T transferred Lenfest's business straight to Comcast, as part of a complex deal in which the Robertses let AT&T bid them out of the Media One cable network. Lenfest learned of the transfer when *Philadelphia Inquirer* reporter Reid Kanaley called him looking for comment. "I won't repeat the language that I used." Ralph had told Lenfest for years "he would own everything in the Philadelphia area."

In a picture frame on the wall in his hallway, Lenfest preserves a typed note he sent Roberts during Comcast's audacious, aborted, but ultimately successful attempt to acquire the assets of Media One in 1997: "You said it was Comcast's destiny to own everybody around you. I no longer doubt it!"

Roberts's scrawled response: "Why don't you get off your high horse and become our partners on a nice quiet friendly pony." Lenfest said he kept the note "so Ralph can't deny he wrote it."

Was that taunting? Intimidation? "He wouldn't have written it," Lenfest said, "if we hadn't been good friends." He added, "But now my saddlebags are full!"

Lenfest said he's determined to give his fortune away while he's still alive. "Hindery did me a favor. I ended up with Comcast stock. It did a lot better than AT&T," which dropped quickly under the weight of all those high-priced deals engineered by Chairman Michael Armstrong.

Did he get out at the right time? "Cable faces very severe competition. Satellite doesn't have to pay for distribution. Satellite doesn't have to pay a percentage of gross revenues to municipalities." And cable operators can no longer extract big "concessions" from programmers, now that cable channels like ESPN have become enormously popular. The programmers "extract a pound of flesh from cable operators—they are very rapacious," Lenfest said.

But then he smiled. "Cable has ways of competing effectively with satellite. Cable has advantage in high-speed modem, in high-definition, in the Internet."

All that is someone else's problem—Brian Roberts's, for example. Lenfest is focused on the dispersal of his estate: He said he wants to donate it all to good causes over the next twenty years. Unlike Ralph, he doesn't believe in dynasties. "In due course, it'll be a major challenge how to give that money away," Lenfest said. To do it, Lenfest has been giving more than $100 away for every $1 Ralph's personal foundation gives out each year.

Lenfest noted part of the Roberts family money is still tied up in company stock. And then he laughed: Ralph "is tighter," Lenfest said. "I'm giving away Ralph's money."

16
DOING UNTO OTHERS

Charity Muscle

Steve Ross, the parking-lot mogul who built the Time Warner entertainment conglomerate, might seem an unlikely patron saint. Lavishing tens of millions in shareholders' money on himself and his inner circle, he narrowly avoided a federal indictment that sent his closest associates to jail.[1] But, then, Ross was generous: He backed up-and-coming movie people like Barbra Streisand and Steven Spielberg, Democratic Party leaders like Jimmy Carter and Bill Clinton, Jewish causes (he was born a Rechnitz, in Brooklyn), and the national cable TV business, which he championed and helped create; and those who received, cherish his memory.

All of this helps explain that annual institution, the United Jewish Appeal–Federation of New York's Steven J. Ross Humanitarian Award Dinner, which gathers cable TV executives yearly in support of Israel and other Jewish causes, and was hosted in 2003 by Ralph and Brian Roberts, whose Comcast had just eclipsed Ross's Time Warner as the nation's biggest cable company. The Robertses include UJA-supported youth leadership programs among the causes they support financially.

But beyond the emphasis on mitzvahs, the evening shone a light on the Robertses' new weight in the entertainment business.

"How powerful are Ralph and Brian Roberts? They got cable's heaviest hitters (including Rupert Murdoch [News Corporation], Dick Parsons [Time Warner] and Sumner Redstone [Viacom]) to ignore the elevated terror alert," passing guard dogs and bomb and metal detectors to join the gala at the Waldorf–Astoria Hotel, reported the *Cablefax* newsletter.[2]

Entertainment included Philadelphia singer Patti LaBelle, who forgot her lines, and New Jersey–bred Comedy Central jokester Jon Stewart (born Jon Liebowitz), who drew protest from a Catholic group when he compared Pope John Paul II in episcopal regalia to "the grand wizard of the Ku Klux Klan."[3] But the highlights of the night were Court TV executives Bob Rose and Henry Schlieff hobbling through a video skit, Schlieff on one leg, Rose apparently missing an arm. Announcing they'd just finished negotiations with the Robertses, Schlieff told Rose, "I don't think we fared that bad!"

They got the biggest laugh. For such events allow the great to poke fun at the greater; and Comcast was indeed squeezing programmers, as it had promised it would, justifying the mergers that had lately made it bigger than Time Warner. That might not be a laughing matter for guests like Murdoch, Parsons, and Redstone, when their attempts to raise costs were parried by Brian and his bean counters. But this gathering of fellow donors could laugh in recognition at those workday struggles and remind this true media elite, or try to, that they could also work together in a larger cause.

Not Giving Much Away

How much does Comcast give back to its community? An economist might ask if it's not enough to employ 60,000 people, sell a service 20 million Americans want, and maintain shares worth more than $50 billion on the stock market. But in real life, corporations, especially those that depend on mass public patronage and government goodwill, are also expected to kick cash back to nonprofit groups, both as a good public relations move and as a good example to others.

Like most big companies, Comcast has set up a foundation, legally based in the tax-shelter capital of Wilmington, Delaware, to funnel that money among the more or less worthy causes that clamor for it. It's not a very large foundation, compared to the size of the company. In 2001—the last year in which Comcast's financial records weren't clouded by deals that either more than doubled the company in size or nearly cut it in half (buying AT&T and reporting a loss in 2002; selling QVC and booking record profits in 2003)—Comcast collected more than $10 billion from its customers, claimed $2 billion in profits, and gave away $2 *million* to charity through the nonprofit Comcast Foundation. The company has given $2 to $3 million away in each of the years from 1999 to 2002.[4] The Comcast–Spectacor Foundation gave away another $1 million, more than half of which was raised, not from Comcast, but from Flyers and Sixers fans, by groups including Flyers' Wives Fight for Lives.[5]

How does that compare with other corporations? The typical big company gives about 1 percent of its profits to nonprofit groups to bolster public relations or further the boss's personal causes, according to the Center on Philanthropy at the University of Indiana.[6] Comcast gave away one-tenth of 1 percent, or $1 for every $1,000 in profits—about one-tenth of what most American companies gave. Comcast's biggest grant, about a quarter of the total, went to a company program that doled out some 659 scholarships of $1,000 each in Comcast's East Coast and Midwest divisions; other big grants supported survivors of the 9/11 terrorist attacks, United Way, and more than thirty smaller organizations.

Ralph Roberts, who has collected over $200 million from Comcast and its shareholders in the years since 1998, has given away around $1 million a year from his and Suzanne's charitable foundation. Brian, not quite as well paid as his dad until recently, and with more mouths to feed, clothe, and educate at home, put up another $200,000 from his and Aileen's foundation in 2001; the total tripled in 2002, according to their charitable tax return. Brian's largest gifts went to the University of Pennsylvania, his alma mater; Ralph's, to Philadelphia's Einstein Hospital foundation. The rest was scattered, mostly in small amounts, to dozens of sports, and Jewish, school, and antipoverty programs in Philadelphia and beyond.

The Robertses have been among the best-paid executives in America since the mid-1990s. Unlike the collectors of two-dozen other late-twentieth-century Philadelphia fortunes, Comcast's founding family has not used a noticeable chunk of its fortune to adorn school or college buildings, or cultural venues, with its name. The Roberts clan sat out the years-long campaign to find name patrons for Philadelphia's new concert hall complex; Sidney Kimmel (Jones New York), canny old investor Ray Perelman, and Comcast phone-company competitor Verizon Corporation got that honor. The Robertses also stepped aside when the Philadelphia Museum of Art rallied big-money donors to finance a proposed move of the fabulous Barnes Foundation's Impressionist masters to a new Center City museum; the Pew (Sunoco), Annenberg (*TV Guide*), and Lenfest (cable TV) families offered to take that expense and labor upon themselves.

Ralph and Suzanne's older daughters are personally active in the city's two most prominent cultural institutions, both of which have long ties to Suzanne's family, the Fleishers. Catherine Roberts Clifton is a director of the Philadelphia Orchestra; she and her husband personally (and quietly) bailed out a series of neighborhood concerts in the summer of 2003 when the money wasn't otherwise available. Lisa Roberts Seltzer has played a roughly similar role at the Philadelphia Museum of Art, donating services from her design business and participating actively on the busy museum board.

In early 2002, it looked for a moment as if the city was going to get a Roberts family charitable memorial—not a fancy building in gleaming Center City, but a highly useful institution in a Philadelphia slum neighborhood, a seeming complement of the century-old Samuel S. Fleisher Art Memorial, the free school that still recalls Suzanne's family in a city where their long-prominent name has mostly vanished from public view. In April, the *Philadelphia Daily News* reported that Brian and Aileen had given $1.5 million to Sister Mary Scullion, the city's best-known crusader for the homeless, toward construction of the $10 million Honickman Roberts Learning Center. (The other donors were the couple named by *Forbes* magazine as the richest in Philadelphia: beer-and-soda baron Harold Honickman and his wife Lynne, an heiress to a suburban real estate fortune).[7] The following winter, a picture of the groundbreaking appeared in the *Jewish Exponent*, with Brian and Aileen, stiff in their winter coats, gamely turning over earth with the Honickmans and Sister Mary.[8]

Nevertheless, two years later, the Roberts name was conspicuously absent from the building they'd promoted. It wasn't the "Honickman Roberts Learning Center," but rather the "Honickman Learning Center and Comcast Technology Labs," which tried to train the homeless for the Information Age. Comcast cash—$1.5 million, spread over ten years—was paying for the computers. Brian and Aileen also gave. Still, Comcast, not the family, would get the credit.

Why doesn't Comcast give more? Corporate leaders "can't do everything," insisted David L. Cohen, the quick, studious, and well-connected Philadelphia lawyer hired by Comcast to oversee the company's public face in 2001. "We have to pick our battles." The company's main obligations are to its shareholders, its employees, and its customers, who are now spread around the country, though Cohen estimated that Comcast gives a relatively high proportion of its donations, around 15 percent, in metro Philadelphia.[9]

Gerry Lenfest laughed when asked why he appears to be giving away far more money than the man who bought him out. For one, he said, he can afford to, having cashed out his stock. Also, he said he doesn't believe in great inherited wealth and prefers to give most of his away before he's dead; he scorned the idea that anyone can use it to form a dynasty, because it's so hard to know if one's descendants can be trusted to follow a donor's wishes.[10]

Comrades

Ralph accomplished something almost unprecedented in Wall Street–dominated corporate America: He passed not just his billionairehood but also the daily control of his publicly traded, investor-owned business down to his son. Of course, the Robertses didn't do all this alone. They rely on driven, focused, loyal deputies, able both to execute and to expand on the boss's orders. To build his team Ralph depended on cold hiring calculations as well as on ties of marriage, blood, and the common life experience of men he chose early to follow and to guide him. Brian has assembled his own successor team; it is a more diverse group, drawn not from the ambitious Jewish immigrant ghetto of his father's time, but from the common national culture of the entertainment business.

In fact, since Dan Aaron, the old newspaperman and refugee from Nazi Germany, had set up Comcast's first operating systems in Mississippi, Comcast operating executives have come and gone with some frequency. Since illness and advancing age hastened Aaron's retirement, the job has passed to a series of hired guns: regional cable chief Dick McCaffery, to whom Brian reported in his Trenton stint; Robert B. Clasen, president of Comcast Cable in the 1980s, who handled acquisitions, programming, and marketing while Brian concentrated on daily operations; Thomas L. Baxter, who ran the cable arm for most of the 1990s; and, most recently, Steve Burke, who, like his boss, Brian, grew up in the cable business, and who makes most of his day-to-day decisions with little interference—a policy Brian follows even more strictly than Ralph.

The Burkes are an upstate–New York, Irish-Catholic version of the Robertses. Burke's father, Daniel, was junior partner with his neighbor Tom Murphy in building a single radio station into the Cap Cities–ABC broadcasting network. His uncle, James, was chief executive of Johnson & Johnson, the drug maker. Burke ran Walt Disney's retailing and European divisions under Michael Eisner before quitting in 1998 to run Comcast's cable arm, moving his wife and five kids to Philadelphia (which he preferrred to California), at the invitation of Brian Roberts's executive headhunter in Philadelphia.[11] Roberts immediately pushed

Burke to the fore, not only as the director of Comcast's tech-themed service expansion, but also as the man who would run Disney—return to
Disney—if Comcast eventually managed to buy it.

Burke was a late addition to the team that had been guiding Brian
for the better part of the 1990s. The financial, legal, and deal-making
part of the group includes co-chief financial officer Lawrence S. Smith,
former chief financial officer of the family-owned Advanta loan company, who joined Comcast in the late 1980s as it was moving from Bala
Cynwyd to Philadelphia (distinguishing himself by flushing the office
goldfish down the toilet when it proved too much bother to transport),
and has since served as Brian's tough-minded reality check; affable John
Alchin, formerly of Britain's National Westminster Bank; and lawyer
Arthur Block and merger specialist Robert Pick—known to some at
Comcast as the team of Block and Pick. Amy L. Banse joined Comcast in
1991 to work on programming deals after leaving the Philadelphia law
firm of Drinker, Biddle & Reath, which has represented Comcast in a
string of mergers. Six years later she was made head of Comcast's new
Programming Investments Department, where her job grew to include
the company's niche programming units—Comcast SportsNet, E! Entertainment Television, Style, The Golf Channel, Outdoor Life Network, the
G4 computer games group, and African American–oriented TV One.

Among the recent additions are David Cohen, former head of Philadelphia's influential Ballard law firm and tireless chief of staff for ex–
Mayor, now Pennsylvania Governor, Ed Rendell, hired to help craft Comcast's message to the public and the Washington power structure; and
Karen Dougherty Buchholz, an energetic and relentlessly organized public relations executive, who helped recruit Cohen after they worked together landing and putting together the Republican National Convention
in Philadelphia. She is part of a power couple: Husband Carl Buchholz
is, like Burke, a major fundraiser for President Bush; Buchholz spent a
year on Bush's staff after the 9/11 attacks and has since been marketing
his government service from his Philadelphia law office.

Empty Nesters, Passive Heirs

At the Elkins Park house where they'd raised five children, Ralph and
Suzanne stayed on longer than some of their friends expected. Elkins
Park, with its solid suburban homes and large Jewish population, recalls
Ralph's native New Rochelle, and it was considered an ingenious and
comfortable place to raise a family when he designed their home ac-
cording to the broad precepts of Frank Lloyd Wright in the late 1950s.

But a generation later, when the kids were gone and the city's suburban fringe was studded with the far more grandiose palaces of Reagan-era millionaires, neighbors and city business leaders cited the Roberts house as a modest, even frugal, castle for a man of Ralph's station.

This time it was Suzanne's turn to take charge. At her bidding, Ralph acquired a quarter-square-mile of farmland in Chester County's Brandywine Valley, not far from estates owned by some of the region's wealthiest families, the du Ponts and the Dorrances (who control Campbell's Soup). Suzanne had stuffed the homes of their early marriage with Colonial American antiques; this time she was determined to create a fitting home for her functional museum pieces, designing from scratch a farmhouse that wouldn't have looked out of place, in a well-scrubbed way, when General George Washington lost the Battle of the Brandywine in the neighborhood back in 1777.

The children, of course, were firmly established in their own lives. If Brian Roberts patterned himself on his dad, learning personal elegance, cutthroat deal making, and bold opportunism from the bottom up, the older children followed Suzanne's old avocations by entering what used to be called the helping professions. Ralph Jr. became a research psychologist in Colorado, where he'd gone to college. The daughters stayed closer to the nest: Lisa has settled in Philadelphia, where she runs her design firm. Catherine, an honors student in city planning at Penn and in public health at Harvard, taught at Tufts for a while, then came home after marrying Anthony Clifton, perhaps the best known of the group in Philadelphia after Brian. The couple were married in 1980 by city Judge Lois Forer at the Roberts home in Elkins Park. A handsome Briton who worked briefly for Comcast, Clifton even showed up in the back of the 1982 annual report, listed below a couple of Fleisher family members as a Comcast "Assistant Treasurer." But Clifton didn't last; he left the company for his own more modest media principality, serving as publisher of a chain of giveaway newspapers, including the *Philadelphia Weekly* and the *South Philadelphia Review*. Brian, Lisa, and Catherine all married bankers, of whom Lisa's spouse, ex–Lehman Brothers executive David Seltzer, was the most professionally successful.

Youngest brother Douglas, like Brian, chose a more aggressive career than the older siblings. He went to law school in Cleveland, served in the Philadelphia prosecutor's and solicitor's offices, and made the news briefly in 1998, when he donated $6,000 to then-Mayor Rendell and made known his interest in getting elected as a city judge. At Ralph's request, Mayor Rendell made a call to Bob Brady, the city's chief Democratic patronage dispenser, party chairman, and U.S. congressman, in an attempt to list Douglas on the ballot. "He's a young guy, and I think he'd be a terrific judge," Rendell told the *Inquirer*, adding, "I've decided not to hold Ralph and Brian against him."[12] But Brady said Doug's pedigree wasn't enough to prevent him from having to run the gauntlet of the city's itchy-palmed Democratic ward leaders prior to gaining endorsement. In the end, Douglas didn't run.

In a 1998 transaction, Brian took over his parents' voting stock, along with his brothers' and sisters' claim on the family company. "Brian's increased ownership . . . demonstrates his continuing commitment to Comcast," which he deserved by having "proved himself time and again," Ralph Roberts said in a public statement.[13] The deal also served "to enable Brian's siblings to diversify their assets without affecting Comcast." Like the old New York railway baron and patriarch Cornelius Vanderbilt, Ralph Roberts saw little reason to attempt an equal division of his business holdings when one son showed interest and talent for the business beyond that of his siblings. If the others feel any discontent, they have kept it to themselves and out of the courts. Indeed, unlike Suzanne, who followed the old Fleisher tradition of spending a lifetime courting publicity for her causes and hobbies, and who delights in naming all eight grandchildren in her public-service announcements for pet charities like the Fox Chase Cancer Center, the Roberts children and their families—Brian's included—have generally shunned personal publicity.

17
FRIENDLY POWERS

Trade Show for George Bush

It was George W. Bush's big show. But even hardened Washington-watchers had the amused impression the 2000 Republican National Convention in Philadelphia was one big ad for Philadelphia's cable TV company.

"Welcome to Comcast Country" read the billboard on I-95, between the airport and the sports hall temporarily retitled the "Comcast First Union Center," ignoring First Union National Bank's longstanding, $40 million "exclusive" naming deal. Comcast's name was everywhere: at each security post and gate, on the lanyards that held convention credentials, at the high-speed e-mail kiosks. Comcast volunteers accompanied delegates to five nights of Comcast-sponsored parties in honor of those members of Congress who happened to write cable TV laws. A glossy hundred-page Comcast convention guide featured more pages about the company than about the future president. Comcast's QVC home-shopping network offered a convention-memorabilia sale. Most important was the Comcast TV and telecom studio, at what was then the First Union Center, through which Comcast controlled, granted, and restricted each politician's precious access to the voters back home.

This was serious branding, for a powerful audience, on an impressive scale. "Comcast has turned the convention into a five-day advertisement," declared Bloomberg LP.[1] "Is this a political convention or a telecommunications trade show?" asked the *New York Times*.[2] "They have sold the name of this convention," proclaimed Larry Sabato, head of the Center for Government Studies at the University of Virginia. Comcast's use of the convention for self-promotion was unprecedented, and probably a trend-setter, according to Sabato: "Four years from now it could be the Tidy Bowl Convention or the Pepsi Challenge Convention," he told the *Inquirer*'s Patricia Horn.[3]

Comcast wasn't the host just in name, but also in fact. Mayor Ed Rendell had naturally looked to the largest company still based in Philadelphia for help landing the Republican show. As the owner of a fabulously wired sports hall and studio, Comcast had something even better than cash to offer the GOP. A marketing enterprise trafficking in images, Comcast knew how to leverage publicity. A multiregional media company with huge national aspirations that depended on government goodwill, Comcast had every reason to make nice with the nation's majority party. So, at the end of July, Republican delegates from Georgetown and Scarsdale, Orange County and Boca Raton, North Dallas and Grosse Pointe—led by half of Congress, including all its leaders, and joined by the nation's lobbying, corporate, and media elite—descended from jet, Metroliner, and oversized car into sweltering South Philadelphia for the week-long, party-studded pep rally that would anoint George W. as prospective president. Their host was, to all appearances, neither the GOP nor even Philadelphia, but the local cable TV company, which pitched its name by means and methods that might have been familiar to anyone who'd watched founder Ralph Roberts hawking ties or golf clubs or Muzak half a century before.

Since Comcast was still only the No. 3 cable company, still mostly unknown in most of the country, for some visitors its self-promotion had a parochial ring to it, more what you might expect in Kansas City or Birmingham than in a big East Coast city. "What is this Comcast?" asked a Southern delegate. "Comsat?" asked Los Angeles Mayor Richard Riordan. "They're everywhere," Riordan said of Comcast. "I'm surprised they're not on the toilet paper."[4]

But who was using whom? If Comcast had slapped its label on the GOP's pep rally, it was making itself an especially useful parasite: "They've got the arena, they've got the cable. They've got the hearts and minds of the grateful Republicans. They may even get to sell t-shirts," chimed in Larry Makinson, Washington political-money expert and head of the Center for Responsive Politics. "This," he concluded, "is how you become a player."[5]

America's Mayor

The Philadelphia region gained more corporate headquarters in the 1990s than New York or Chicago, Boston or Atlanta, Denver or Phoenix.[6] But most went to the suburbs, not downtown. Comcast has a special importance in Philadelphia because it is the biggest of a handful of large companies that, over the past generation, have grown and thrived while based in Center City, which is otherwise remarkable as an industrial museum. The neighborhood hosts, besides a fine collection of restaurants and a surprising number of rehabbed fancy apartment buildings, the usual downtown assemblage of law firms, out-of-town retailers, and financial branch offices. Still advertised on the walls, doors, and top floors of many of its high-rise towers are the names of railroads, banks, stores, and factories that no longer exist but have not been replaced by anything more dynamic.

Comcast's clout was reinforced by the character of the city's elected leaders, especially Ed Rendell, mayor of Philadelphia for most of the 1990s. Rendell is a Democrat—was, in fact, the chairman of the national Democratic Party during its 2000 national campaign—but like other Clinton-era Democrats he was above all a practical politician. If his critics dwelled on how he could be moody, bullying, reluctant to fight the really tough battles, his admirers—from brainy Center City lawyers to loyal rowhouse committeemen and even the suburbanites who have since made him Pennsylvania's governor—saw Rendell as a tower of smart and relentless enthusiasm, at least for those fights he deemed winnable. So when he saw a chance to lure the Republicans to his city, filling tens of thousands of taxpayer-subsidized hotel rooms with an army of influential and free-spending visitors, the mayor strapped on an elephant necktie, rallied such corporate leadership as remained in Center City, and managed, against stiff competition, to bring that really lucrative party home.

Like Ralph Roberts a generation before him, Ed Rendell was an ambitious Jewish New Yorker who came to Philadelphia as a teenager and made useful friends quickly. Also like Ralph, he graduated from the University of Pennsylvania, married a polished local girl, and never left. Similarly, too, Rendell was a born salesman, who drew money to himself, though Rendell's focus was political power, rather than business and the family fortune. Running short of campaign funds in his first bid for elective office (as Philadelphia's top prosecutor), the ambitious young candidate had been rejected by a string of loan officers from the big Center City banks. But Rendell had kept knocking until he found a suburban bank willing to write him a "home improvement loan," on the grounds he'd likely move to a nicer house if he won the race. He won.

Rendell spent most of his two terms as the genuinely popular and business-friendly mayor of a Philadelphia that, despite its proliferation of pricey restaurants and fancy but underused hotels, was still losing what remained of its corporate elite. Back when Ralph Roberts was starting his cable business, the men who had run the South Broad Street banks and brokerages, the Market Street department stores, the specialty manufacturers in North and West and Northeast Philadelphia, even the railroads descended from the powerful old Pennsylvania line and its competitors, might be found relaxing together at the Union League, or the Philadelphia Manufacturers or Merion Cricket clubs. But no more. Their successors, mostly hired men who'd grown up somewhere else, spent the boom years of the 1980s and 1990s cashing in stock options and selling out their companies to the highest bidders. Unlike Ralph Roberts, who was founding a dynasty, they cared little about what might happen after they were gone. When they had the chance, they sold out, took the money, and ran.

Successful at raising money, facing down city unions, and generating a positive buzz, Rendell was flummoxed by the continued disintegration of Philadelphia's corporate core. What power did he have, really? He wasn't the sort of mayor to lie down in front of bulldozers. And he wasn't about to do much about the real problems of high business and wage taxes, outrageous parking costs, and crummy public schools that drove big companies to the suburbs. Rendell called himself the ultimate frat boy; he was a pushy, glad-handing realist who did his best to get along with the powers that be, in hopes of persuading them to intervene from time to time to his city's advantage. Rendell could be tough on city trash men; but his strategy with big business was lots of carrot, not much stick.

He had tried flattering the companies that were taking over Philadelphia—even agreeing to support North Carolina–based First Union Corporation's takeover of Philadelphia's last big bank, Philadelphia National (CoreStates), which had funded Comcast's first deals, in exchange for the hope, not even a promise, that the bank might move some low-wage jobs to town. It didn't: First Union cut 10,000 jobs in the Philadelphia area, and Center City took the worst hit as the company moved back-office jobs to Reading and Wilmington.[7]

Out beyond City Line Avenue, in the booming corridor from Princeton to Wilmington, MBNA was becoming the world's largest credit-card company; Vanguard, the top mutual fund; Merck, the leading drug maker. But the men who ran these giants usually took little notice of

Philadelphia; their businesses were national, even global, and home was a pleasant green neighborhood, central to the office, the turnpike, and the airport; for many of them Philadelphia was, like Valley Forge or Longwood Gardens, a regional attraction: the high-rise/restaurant/cultural district, and the stadium-bend in I-95, surrounded by a slum.

Comcast, by comparison, had become more urban, more Philadelphia, as it gained wealth and influence. Having won their long, patient struggle to wire the more lucrative parts of the city, the Robertses had moved their headquarters downtown as promised. True, there was still no Comcast tower; the company had rejected many attempts to build itself a distinctive headquarters, preferring cast-off space vacated by the vanished banks. Yet Comcast bought better credibility when it purchased the city's indoor sports teams, building the new arena in South Philadelphia instead of Delaware County or (shudder) South Jersey; and when, continuing to grow, it added modest numbers of jobs and contributed to members of city council, showing the loyalty that had vanished elsewhere in corporate Philadelphia. That's why Comcast was so important to politicians like Rendell, and why they aggressively defended the company, even against consumers' best interests, when would-be competitors came knocking; and it made it that much easier for Comcast to play so prominent a role in Philadelphia's convention plans.

Rendell had launched a modest attempt to land the 1996 presidential conventions. His street-smart but boardroom-savvy lawyer act had won the city a certain respect; beyond City Line Avenue, his image far outshone such bumbling predecessors as nasty ex-cop Frank Rizzo or aloof Baptist deacon Wilson Goode. Al Gore had even called Rendell "America's Mayor" and touted him as a role model. Still, in the 1996 convention bake-off Philadelphia hadn't won serious consideration, partly because of the city's lack of hotel rooms. A spate of building and conversions, including hefty tax breaks for Marriott Corporation, took care of that problem. Rendell also assembled a powerful team: For his convention committee the mayor secured as cochairmen David F. Girard–di Carlo and David L. Cohen. Girard–di Carlo was chairman of Blank, Rome, Comiskey & McCauley, a Philadelphia-based, nationally prominent law firm so powerful and effective at acquiring public business from the Republicans who mostly ran the state that it was known in Harrisburg, the state capital, as "Blank Check." Cohen was chairman of Rendell's own Democratic-leaning firm, Ballard, Spahr, Andrews & Ingersol. Cohen had served as Rendell's first mayoral chief of staff and negotiated Philadelphia's fifteen-year franchise agreement with Comcast—a document so favorable to the company that the city's official consumer advocate, Lance Haver, said it was no wonder the grateful company rewarded Cohen with an executive job. In fact, Cohen was the man most often credited with being the real brains of the Rendell administration.

The third cochair was Comcast's own Brian L. Roberts, who brought physical assets and personal connections all his own. "The 2000 convention could have been held at the Philadelphia Convention Center, which had several thousand seats more than the First Union–Comcast Center," noted syndicated columnist Robert Novak. Indeed, the downtown center had been built, at considerable public expense and lasting inconvenience, to house just such an event as a political convention, but it never has. What Comcast did promise was "cable coverage of the proceedings free of charge, saving the party heavy fees," according to Novak. "The deal was brokered by Ken Duberstein, a prominent Republican insider who is also a Comcast lobbyist." Duberstein had been Ronald Reagan's chief of staff; there could be no better evidence that Comcast was playing in Washington's big leagues.[8]

Not the News

"We're not a news organization, and we don't have any real aspirations that way," Brian Roberts told TV interviewer Neil Cavuto as the convention opened.[9]

It was a telling admission. At the Bush convention, Comcast exercised unprecedented control over the three-hour nightly convention feed to 70 million American homes served by a consortium of big cable companies. As *Crain Electronic Media* reported, "Comcast saved an alcove just off the podium for itself, where it could pull politicians off the floor for exclusive interviews. No other media company was allowed access or permitted to have a similar set-up." But Comcast wasn't about to use its media dominance to ask any delegates hardball questions or help them make up their minds about who ought to be running America. To interview the nation's most powerful politicians, Comcast lined up, among others, two GOP operatives, Suzi DeFrancis and Bill Palatucci, along with Lynn Doyle, anchor for Comcast's nascent news operation and wife of Comcast East Coast cable chief Michael Doyle.

No wonder, as Bloomberg LP noted, that in Comcast's coverage, "issues such as gun control and abortion are avoided in favor of discussions about 'excitement' on the convention floor." Comcast cable chief

Steve Burke told Bloomberg LP, "We felt it was an important thing for the convention to make sure the cable industry gave the kind of coverage that the convention wanted." Americans were in no danger of learning what was really going on from Comcast.[10]

Comcast had a message of its own for its politician–visitors: The company, facing pressure from Republicans as well as Democrats to limit its rates, improve its service, promote Internet competition on its lines, and submit to new regulation, was making friends of the men and women who could keep government off its back and help preserve its monopolies. For Comcast's Burke, at the nominating convention for the job of most powerful man in the world, "the most important message is, we function better as a deregulated industry, that's our primary message."[11]

Senator Lott's Constituents

Since Philadelphia is old enough and rich enough to have attracted its share of impressive buildings, but not vital enough to have knocked them all down again for something bigger, the city has preserved some of the quaintest semipublic indoor spaces in America. Most of them were in use at some point during the week of the GOP convention, as corporate America honored its friends in the dominant party. And every night the convention lasted, at least one of these spaces was used by Comcast to fete some of its particular friends in high places.

"I view you as constituents," U.S. Senate leader Trent Lott, R-Miss., told the Robertses with a big smile, at the party they hosted for him on the first night of the convention. Lott's affair was at the Franklin Institute (in the same hall where Brian had met his wife at a fundraiser years before).[12] Ralph and Brian Roberts had never lived south of Philadelphia, but they were indeed important supporters of the powerful ex-segregationist. Ralph had started his cable empire making friends at the lower levels of the Mississippi political machinery in towns like Tupelo and Meridian. Over the years Comcast had proved such an astute corporate citizen that by the 1990s Lott was consulting them on cable legislation, and cashing their checks.

The same day as Lott's party, Comcast had helped host a fundraiser for U.S. Rep. Robert Ehrlich, R-Md., at the Union League Club; two days later, Comcast would entertain Ehrlich at the University of Pennsylvania's Museum of Archaeology and Anthropology.

On Tuesday, August 1, it was off to West Philadelphia, to the studio where ageless Dick Clark had once hosted the enormously popular *American Bandstand* dance program, featuring West Philly girls and boys who set the fashion and the beat for the nation under President Eisenhower in television's "golden age." The guest at the party, co-hosted by AT&T and the Weather Channel, along with Comcast, was U.S. Rep. Tom Bliley, R-Va., retiring head of the House commerce committee. Elsewhere Tuesday, Comcast helped honor Bliley's prospective successors, Billy Tauzin, R-La., and Mike Oxley, R-Ohio, with a Mardi Gras–themed party; for J. C. Watts, R-Okla., a rare black Republican, there was Philadelphia's own smooth-voiced singer Teddy Prendergrass.

And on Wednesday it was back to West Philadelphia and the gala for Ehrlich at Penn's museum, where 200 guests gathered on Comcast's dime, eating crab cakes amid the treasure troves of Chinese dragons, Egyptian mummies, and Mesopotamian idols.

"There's no connection whatsoever between the fact [Ehrlich] serves on the Commerce Committee and the fact we're having this reception for him," Comcast spokesman David Nevins told the Associated Press.[13] "If someone thinks Comcast is trying to buy votes, this is a very cumbersome way of going about it," Ehrlich told a *Baltimore Sun* columnist.[14] When the AP pressed the congressman on just what he was giving Comcast for its support, he replied, "You must have a low opinion of politicians."[15] Ehrlich has since been elected Maryland's governor; Comcast is now the biggest cable company in Maryland and the dominant sports media purveyor in the state; and Ehrlich's wife is a Comcast lobbyist.

You didn't have to prove exactly what Comcast was getting in return to know the company was getting something out of these parties and sponsorships. "Comcast will have an embarrassment of riches in terms of contacts," predicted the University of Virginia's Larry Sabato.[16] "This was a good opportunity to increase the positive awareness of Comcast," offered Brian Roberts at the time, "and I believe we accomplished that."[17]

Two Faces on K Street

It was one of those Washington things: The broadcast industry was Republican, more or less, and their upstart opponents, the cable guys, tended to be Democrats. It had been that way for half a century, since that prominent young Pennsylvania Democrat, Milton Jerrold Shapp, had staffed his seminal cable equipment and finance company from the ranks of the ambitious young socialists who had followed President Roosevelt's left-wing Secretary of Agriculture Henry A. Wallace into the federal government and back out again. Individual cable operators were conservative Republicans often enough, especially as the industry started to prosper; but the industry's Washington lobbyists were more often Ds than Rs, even as cable entered the media mainstream in the 1990s.

It looked like more of the same, in 2001, when Comcast hired David L. Cohen, Ed Rendell's strong right arm during his first and most successful term as Democratic mayor of Philadelphia, to be executive vice president in charge of lobbying and public affairs. Rendell, of course, after chairing the national Democratic Party in 2000, went on two years later to be elected Pennsylvania's first more-or-less liberal Democratic governor since Milton Shapp, thirty-two years before. But Cohen, like Rendell, engages in practical politics. Brian Roberts said he wanted Cohen for his brains, which he'd had a chance to observe at close range on the Republican National Convention host committee, at least as much as for his connections. But Cohen quickly set to work building Comcast an impressive network of top-drawer GOP lobbyists, putting the company more in line with Republican-dominated Washington.

"There has been a perception that the cable industry in general, and maybe Comcast in particular, has had stronger Democratic ties than Republican ties," Cohen told *Roll Call*, the newspaper that covers Congress. "You have to have a Republican as well as a Democratic face."[18]

By the time Cohen joined Comcast, taking charge of a small lobbying staff anchored by ex-FCC economist James Coltharp, the company had already purchased the services of at least one prominent Republican, Ken Duberstein, the former Reagan aide, as an outside lobbyist. In buying part of AT&T in 2002, Comcast also picked up such GOP-connected outside lobbyists as Lorine Card, sister-in-law of George W. Bush's chief of staff, Andrew Card.

That winter, Cohen went into full-throttle hire-more-Republicans mode when he lured Kerry Knott, former aide to House GOP leader Dick Armey, R-Tex., away from Microsoft, where Knott's duties had included trying to shut off a federal government antitrust probe and building the kind of lobbying muscle the software giant wanted, to prevent such a probe from happening again. The move also recognized the "K Street strategy" of Armey and other Republican activists, who were pressuring companies to drop Democratic lobbyists and hire Republicans if they expected to do more business with the party that ran Washington.

Comcast gained more Republican-friendly faces through the spring of 2003. Brian Kelly, former head lobbyist at the GOP-oriented National Association of Broadcasters (and for Walt Disney Company), signed in May. In June, Comcast announced the hiring of Jessica Wallace, Republican counsel on the House commerce committee, which handled cable, TV, and satellite issues. Her old boss, Rep. Billy Tauzin, R-La., announced he "could not be more proud of her." "Our Subcommittee's loss is Comcast's big gain," added Rep. Fred Upton, R-Mich., chairman of the Commerce Subcommittee on Telecommunications and the Internet, in a press release.

Wallace's hiring was balanced a bit by the near-simultaneous signing of Melissa Maxfield, who had headed a political action committee for Senate Democratic leader Tom Daschle. He also issued a press release: Daschle said proudly, "Comcast will benefit greatly from the rare combination of intelligence, energy and ability" that made Maxwell "an essential part of my political family." Knott delicately assured the GOP that Comcast wasn't going overboard by retaining a Democrat: Though "Melissa knows virtually every Senate Democrat," he noted, "her relationships span both houses and both parties."

In December, Comcast added star power to its all-pro lineup by bagging Pentagon spokeswoman Victoria "Torie" Clarke, a onetime cable-industry lobbyist well known to TV viewers for presenting the Bush administration's line on Iraq. (Bush-haters had pounced on her pronouncement she was "not losing any sleep" over Iraqi civilians killed by a failed air strike on Saddam Hussein.[19]) Clarke's previous resume wasn't military, but political; it included not only service to the elder President Bush, but also a term with one of cable TV's most persistent critics, Sen. John McCain, R-Ariz. She "enjoys a sterling national reputation as a trusted communicator," Brian Roberts announced. "We are delighted that she has agreed to join the Comcast family."[20]

Comcast was also making obvious progress with Washington regulators. President Bush had tapped Michael Powell, son of popular Secretary of State Colin Powell, as his Federal Communications Commission chairman, and the younger Powell had given cable the benefit of many doubts on its promises that the technologies (and mergers) embraced by Comcast and other leaders of the cable cartel would promote competition without any need for government interference. At the cable industry's national conclave in Chicago that June, Powell went so far as to take charge of an oversize remote control and offer a public demonstration of Comcast's high-definition TV service.

"It's inappropriate for the chairman to go out and plug this," Jeffrey Chester, head of the Center for Digital Democracy, a liberal public-policy group, told the *Philadelphia Inquirer*'s Akweli Parker, who broke the story.[21] "The public should be wary when Michael Powell comes and acts as a PR prop." Powell declined comment; his spokesman called the chairman's actions appropriate, and added that Powell's practice of traveling to cable shows at the industry's invitation, and with industry cash, saved taxpayers money, Parker reported.

A convenient federal court decision in 2000 had limited the remaining powers of cities and towns to regulate the cable industry, but Comcast still found it expedient to hire well-connected lobbyists to make friends in state capitals. In Maryland, Comcast chose the state's best-known political persuader, Bruce Bereano. Never mind that, in 1994, Bereano had been convicted on federal mail-fraud charges, for falsely billing clients for illegal campaign contributions, and disbarred as a lawyer. Bereano's persistence and clout were such that he was able to continue lobbying legislators by phone from the halfway house where he served his five-month sentence.

In 2002, Bereano lined up behind Comcast's favorite Maryland congressman, Robert Erlich, working hard to help elect him governor. Comcast then retained Bereano—like Ehrlich's wife, Kendel—as a lobbyist in Annapolis, the state capital. But the following summer Bereano went too far. First, on July 1, his lobbying license was suspended by the state ethics board for accepting percentage commissions instead of retainer fees from another client—a no-no under state law. Apparently undaunted, Bereano arranged for a 100-foot sailing schooner to carry Maryland legislators at the National Conference of State Legislators around San Francisco Bay later that month, on behalf of his client, Comcast, during hours when they were supposed to be attending working sessions. But the *Baltimore Sun* got wind of the trip and started calling embarrassed legislators, who said they had no plans to join Bereano. Pressed publicly by Comcast, Bereano postponed the trip until post-conference hours, then canceled it. That wasn't enough to save Bereano's Comcast contract. "We no longer have a need for his services," spokesman David Nevins told the *Sun*, declining further comment.[22]

Wiring the Capital

When the *Washington Post* commented, a month after the Republican convention, that "Comcast wants to cast a very long shadow over Washington," it wasn't talking about lobbyists or political contributions. Having lost the late 1990s merger sweepstakes against AT&T's free-spending Michael Armstrong, Comcast acquired a string of systems AT&T didn't want, including the woeful franchise for Washington itself, as well as the neighboring middle-class and blue-collar suburbs. A company that serves members of Congress and staffers, and serves them badly, runs the risk of alienating people who can make its life miserable. But by buying the District of Columbia franchise, which already suffered a reputation for service glitches and failed promises to improve, Comcast hoped to distinguish itself for better quality, not least by offering a long menu of the same new Internet and video services it was battling to keep unregulated. Comcast aspired "to put our best foot forward here so that the regulators, lawmakers and customers see us in the best light," Brian Roberts said.[23]

The Washington-area franchises also fit neatly with those Comcast already held just to the north, in suburban Baltimore and the Philadelphia area. The combination would create a solid Comcast territory across the mid-Atlantic, from suburban New York to northern Virginia, giving the company the power to sell advertising and consolidate local programming costs. Local programming still meant sports, mostly. Barred by league rules from acquiring more pro basketball or hockey teams, Comcast purchased Home Team Sports, which broadcast the Washington Wizards and Capitals (part-property of AOL Time Warner) from Sumner Redstone's Viacom. Separately, Comcast purchased three minor-league Maryland baseball teams and bought naming rights at the University of Maryland's new indoor arena for $25 million.

A generation earlier, well-connected Washington entrepreneur Jack Kent Cooke had joined his early cable franchise to control of the Washington Redskins and parlayed it into leverage among Washington's elite. But Cooke had many interests. In Washington, Comcast was more than his heir. The nation's capital was another market, its residents another satellite in Comcast's growing constellation.

Lights Out

Even highly placed politicians can't always help their corporate friends the way they want to when those friends aren't all that popular. In early 2004, Governor Rendell proposed a tax break for a new, sixty-story Philadelphia tower that would give Comcast a landmark home befitting its status as the biggest company based in Philadelphia. What would other taxpayers get in return? The company wasn't promising, but Rendell and the project's developer insisted, on Comcast's behalf, that forgoing millions in state and local taxes could eventually persuade Comcast to move additional jobs downtown. Rendell's wishful thinking had failed to deliver corporate employment on earlier occasions, notably the CoreStates merger; and even the governor's hand-picked successor as Philadelphia's mayor, John Street, resisted at first, citing the blow to the city's underfunded budget. Rendell was able to bring Street in line, but he was unable to budge the city's downtown landlords. Worried about rising vacancy rates and a lack of new tenants, and fearful of what all the cheaper taxpayer-subsidized new space (which Comcast alone wouldn't fill) would do to local rents when it was offered on the market, Center City owners and managers howled. Another group that wasn't falling over itself to please the cable company was the state legislature, where the Republicans in charge had grown accustomed to curbing the Democratic governor's enthusiasms. So even if Rendell was sincere in identifying Comcast's interests with the general good, a powerful bloc of conservative business interests wasn't feeling so generous.

The building owners were loud and public in their opposition. In June, they came up with an ingenious protest: They turned off the lights atop the city's tallest buildings, darkening the skyline, and dramatically illustrating the threat they said the Comcast tower posed to Center City. In Harrisburg, the tax break proposal stalled. And Comcast pushed back its high-rise dream and began preparing to rent office space elsewhere, at least for the time being.[24]

BEATING MA BELL

King and Kingmaker

Sure, dad gave him the job, and his inheritance of deal making, debt, and digital dreams. But it was Brian Roberts, with his father's blessing and the guidance of his own lieutenants, who led the surprise-attack mergers that turned Comcast into America's largest cable TV company and its fastest-growing Internet service. By the time Brian's company absorbed AT&T Broadband and dropped that name into history's garbage, in 2002, even the industry's irreverent senior statesmen were proclaiming that Brian had come into his own. "He's no longer the crown prince. He's the king," said Dick Parsons, head of AOL Time Warner.[1] "King of the world," affirmed John Malone, after Roberts acquired the vast and ramshackle cable business Malone had built, from its next owner, AT&T, at a discount.[2]

For twenty years the nation's biggest companies—GE and GM, AT&T and American Express, Time Warner and Westinghouse—had been wiring money into pay TV in hopes they could set up lucrative toll-booths where the entertainment, computer, and communications industries crossed. Companies whose shares are traded every day on the stock exchanges need steadily rising profits. If not, their share price goes down, and some stronger company takes them over. That makes corporate ownership sound like a poor fit for cable TV, given the wide swings in the prosperity and perceived prospects of cable companies, as measured by their changing stock values and deal prices. But that's true of many businesses besides cable. The dream of a wildly profitable convergence among Hollywood programming, Silicon Valley computing, and Main Street television-watching drew the big players in; the handful of large operators, led by AT&T and Time Warner, absorbed most of the family firms that had strung and dug the national cable network.

But in its latest merger, Comcast, the ultimate family cable company, had outdone them all. AT&T was a member of that most exclusive of corporate clubs, the thirty-member Dow-Jones Industrial stocks, whose collective rise and fall measures the health of American investments. So, for that matter, was Comcast's next target, the Walt Disney Company. It seemed as if mainstream corporate America, with its hired-gun executives, relentless investors, and quarterly profit taking, couldn't quite cope with this time-tested Roberts dynasty that really, really wanted to be in the cable business.

Strong-armed

Michael Armstrong was supposed to save AT&T. A systems engineer by training, he had spent thirty years soldiering up through global management at IBM, then five years running the Hughes Electronics unit of General Motors, where he oversaw what AT&T later bragged, on the occasion of his hiring, was "the most successful consumer-electronics introduction" ever: the DirecTV satellite network, the closest thing cable has to competition.[3]

Armstrong had the ideal resume for a savior of corporate America. From IBM to GM to AT&T, with time served on corporate and college boards and federal government committees—Armstrong had the corporate background you'd expect for a new head of Ma Bell, which, despite its government-ordered breakup of twenty years before and despite price competition on all sides, still counted 80 million customers and $50 billion in yearly sales, almost $500 for every household in the United States. Microsoft had invested a billion dollars in Comcast in the late 1990s; but it invested $5 billion in AT&T. AT&T's existing resources and powerful cash flow all made for quite a war chest, if Armstrong chose to buy the businesses he thought AT&T needed and didn't have time to build on its own.

But AT&T also carried a curse. Like Rockefeller's Standard Oil Company, like Citibank and IBM and Microsoft, AT&T had been a political symbol for the business power that Americans regard with emphatic ambivalence. AT&T had been a regulated monopoly, extending phone lines even where they weren't profitable in exchange for government-sanctioned profits on basic services. But in the 1970s the Justice Department broke off the regional Bell operating companies, in hopes of stimulating competition that could lead to more and better services, and lower rates. In fact, the regional Bells controlled the wires into most homes and businesses; they also enjoyed deep pockets and plenty of cash flow from monthly billings. Bill Gates was seen to be leveling the playing field a bit when he backed the cable companies against the Bells in the late 1990s.

AT&T, competing with its own progeny as well as new long-distance phone competitors like MCI and Sprint, cut prices and profit margins and redoubled its lobbying efforts to compete. And it cast about for a strategy in this strange new digital data world, which threatened to make old wire-based companies obsolete. AT&T no longer enjoyed market power, yet it felt its movement still restricted by its reputation as a once and potential monopoly. "To my knowledge, no other American company has been put through such a regulatory and court-ordered 'wringer,'" said Walter Elisha, head of AT&T's search committee, when he announced Armstrong's hiring.[4]

Armstrong's solution was to buy both assets and goodwill. His AT&T would cover all the bases—cable and wireless as well as long-distance phones. Armstrong spent more than $100 billion on creating what was briefly the nation's biggest cable TV business. And, to please regulators, he agreed he'd use it to carry other companies' Internet services, angering the same cable companies he'd have to rely on if cable-by-phone and other services Armstrong was counting on were to become reality anytime soon.[5] "Over the next five years, we're going to see a breathtaking transformation of this [telecom] industry, and I'm delighted to join AT&T—the company I believe is going to lead that transformation," Armstrong announced. "This company has the experience, the people, the technology, and most important, the commitment to get the job done. Personally and professionally, I can't wait to get on with it."[6]

More for Less

Among the first people Armstrong got on it with, after firing nearly 20,000 AT&T workers and making the usual noises Wall Street wanted to hear about streamlining and realigning business units, was John Malone at TCI. Among its many other investments, TCI was the biggest partner (along with Comcast and other cable companies) in Teleport, one of a number of new companies that offered fiber-optic phone and data services to business clients in metro areas, competing with regional Bells. Armstrong bought Teleport for $13 billion in AT&T stock; that made cable in general, and Malone in particular, significant AT&T shareholders. Then Armstrong started talking to Malone about buying all of TCI. A big base of cable customers, Armstrong believed, would give his company an edge in selling phone and Internet services. In June 1998, Armstrong paid $48 billion for TCI and its affiliates, and AT&T was suddenly the nation's biggest cable company. Malone got to keep control of the company's Liberty Media programming affiliate, while getting rid of not just the tiresome, largely unimproved TCI cable systems, but also, with them, the lobbying, legal, and public relations headaches that came with being the biggest man in cable.[7]

If TCI could be bought, everyone in cable was for sale. Comcast wanted to be among the buyers. In the summer of 1998, Ralph Roberts invited Glenn Jones, owner of Denver-based Jones Intercable, to breakfast, and informed him he was taking over Jones's company: "We want to issue a press release, but I didn't want to do it without you being involved," Jones recounted.[8] And then in April 1999, Brian Roberts briefly one-upped Armstrong, announcing that Comcast would pay $58 billion for Media One Corporation, the cable business that had been owned by Denver-based AT&T spin-off U.S. West, which AT&T coveted. This was Brian's new empire building, not Ralph's old bargain hunting: Comcast was offering to pay more, for less, than AT&T had paid for TCI just ten months earlier.

But Armstrong, with the approval of Microsoft and Media One's largest shareholder, cable pioneer Amos Hostetter, came back with a higher offer of $63 billion. Brian folded, in exchange for $1.5 billion, plus purchase rights to certain cable properties, like Lenfest's, which gave Comcast something it had been unable to purchase outright: a lock on the Philadelphia market. Armstrong was now Number One, if size mattered. Now all he had to do was show he could make it pay.

Under Malone, TCI had been notorious for annoying consumers and deferring system upgrades. Armstrong's aggressive borrowing to fund acquisitions had his investors demanding immediate success. But AT&T couldn't deliver faster, better, more sophisticated Internet and data and video-on-demand services immediately—certainly not on TCI's frayed network. It would take time and money to improve the system and develop the markets. With the stock market falling through 2000, AT&T

lost more than half its market value. Armstrong gave up. In October, a year and a half after his costly Media One victory, Armstrong announced plans for breaking up AT&T and its affiliates into five companies: the slow-growing phone service, the competitive business-services unit, wireless, cable, and media programming—John Malone's Liberty Media.

AT&T's cable arm had been successful in one respect: It had served as a cash cow, throwing off as much as a dime on every dollar of sales to Armstrong's parent company, according to its financial reports. That meant its gross profit margins were less—20 cents of every dollar, compared to 30 to 40 cents for Comcast—but a buyer who could afford to stop skimming that extra cash would show rapid financial improvement. Armstrong intended that improvement would accrue to AT&T shareholders through a new, independent cable company, AT&T Broadband. But Brian Roberts didn't want to let that happen. He saw an opportunity in Armstrong's weakness, and he wanted to get there before anyone else.[9]

"Don't You Dare Give Up"

Roberts broached the matter with Armstrong at industry gatherings. They discussed it over dinner. But Armstrong didn't take seriously the idea of Comcast's owning a big piece of AT&T, the piece he had spent so many billions to build.

Early in 2001, Comcast put the merger department at its longtime New York law firm Davis, Polk & Wardwell to work drafting a plan for a hostile takeover of AT&T Broadband. Brian Roberts and Michael Armstrong had spoken informally—they both agreed size mattered—but Armstrong continued work on the AT&T break-up as if Comcast weren't a factor. He certainly wasn't disposed to let Comcast go on record as the buyer, as if it were America Online taking over Time Warner, just because crazily inflated Internet stock prices made it temporarily possible for the smaller company to buy the big one.

On July 3, AT&T published a formal spin-off plan, which made no mention of a possible cable sale to Comcast or anyone else. Brian was nearly frantic; he missed his father's counsel. Ralph was vacationing in Europe, and Brian had an awful time sending Ralph papers to review. Finally, Brian shifted his own vacation to meet Ralph on Rittenhouse Square over the July 4 holiday. Ralph gave his blessing. Thus fortified, on Sunday, July 8, Brian struck.

Michael Armstrong was relaxing with his grandchildren in Nantucket when Comcast's Steve Burke called to inform him Comcast was launching a hostile $44 billion bid. Armstrong didn't want to deal with it just then; Comcast sent out press releases and news announcements and forced him to take it seriously. "I'm certain it came as a complete surprise when we put in the unsolicited offer," Comcast financial officer John R. Alchin said months later. "But business is business."[10] It was, general counsel Arthur R. Block later bragged, a historic first: "a hostile tender offer for a division of a company."[11] Armstrong wasn't the only person surprised to see a Philadelphia cable company taking on a target bigger than itself. But people who'd dealt with Ralph and Brian laughed at the notion they'd just walk away. "The Roberts[es] don't take 'no' for an answer. They repeatedly don't take 'no' for an answer," former FCC chairman Reed Hundt told the *Wall Street Journal*.[12]

How badly did Comcast want AT&T? Again and again, Ralph Roberts had let investors and even deals slip away rather than give up control over his company. At the yearly media/technology retreat hosted by Microsoft cofounder Paul Allen in Sun Valley earlier in July, Viacom chief Sumner Redstone told Brian, "Don't you dare give up your voting stock, because that's what defines Comcast. . . . That's what makes it special and makes it different. Investors can buy and sell. That's how they vote." Brian treasured Redstone's comment among his "poignant" memories.[13] Yet Roberts family control was a sticking point for AT&T's many public investors. Corporate reformer Nell Minow of the Corporate Library in Washington, D.C., called the Comcast board a "control freak" for trying to preserve the Robertses' voting control at the expense of other investors.[14] Ralph and, now, Brian controlled more than 80 percent of Comcast voting shares despite owning less than 3 percent of the company. It's not a unique arrangement—the Nike sportswear company is run that way, and the owners of the *New York Times*, Dow Jones, and the *Washington Post* have similar arrangements—but AT&T and its many pension and mutual fund investors forced Brian to cut his voting strength to one-third. The compromise appeared to reduce Brian's role, but actually left him firmly in charge, especially with a provision that effectively guaranteed Brian the combined companies' top job for ten years unless he voted against himself. At New York's St. Regis Hotel, where Brian, Steve Burke, and their team pitched camp to begin their campaign, Ralph shrugged: "You give a little to get a little."[15]

The offer halted AT&T's spin-off proposal and drew counter-bids from Time Warner and Walt Disney Company, among others. Roberts kept calling, but talks went nowhere until Comcast finally agreed to AT&T's demand for secret negotiations in September. Disney dropped out, but Atlanta-based Cox Communications joined Comcast and Time Warner in bidding before the November deadline. On December 7,

AT&T's board, advised by merger-defense specialist Martin Lipton, rejected all the bids. But general counsel Charles Noski called Brian, reaching him at QVC's Christmas party, and begged for a sweetener. Roberts and Burke rode to New York for lunch with Armstrong and Burke that Saturday, racing back just in time for the Philadelphia social event of the year, the black-tie opening of the giant glass Quonset hut of the Kimmel Center for the Performing Arts. Comcast bid again, this time offering $50 billion in stock plus $20 billion in AT&T debt; Time Warner and Cox bid again, too. On Wednesday, December 17, Roberts, his father, and his top managers took the train up to New York, where they set up at the St. Regis Hotel to wait for final approval. It came at 5:45 that evening: "Are you doing anything tonight?" Armstrong asked Roberts. It was official.[16]

The Bigger They Come

The biggest media deal of the Internet bubble, the one that eclipsed the Media One bidding war and the TCI sale, was America Online's 1999 purchase of Time Warner for $110 billion. The older company's publishing, programming, and cable empire, including HBO and CNN, grossed as much in three months as AOL's entire Internet services in a year. But despite Time Warner's greater size, AOL's status as the only Internet-based company to make piles of money from regular subscribers had lifted its stock to such heights that it was able to purchase Time Warner, instead of the other way around. Upstart AOL, which, like cable TV, enjoyed a mass market of steady users while making more money than it liked to admit from pornography, had taken over one of the major media programmers. The New Economy, the Internet and telecommunications companies that were supposed to wipe out traditional business, seemed to have triumphed.

It was premature. The AOL–Time Warner deal would eventually serve as a hallmark for the excess of that bright, short, speculative moment. Comcast's purchase of AT&T Broadband, a larger company with smaller profits and bigger problems, proved a sounder deal, at least in its first few years.

Like many bold and desperate moves, Comcast's was in part defensive. Rupert Murdoch—baron of the English-speaking press, owner of the Fox TV networks, and patron of European satellite TV—had been laboring for years to break cable's hold on U.S. pay TV by taking over a U.S. satellite TV system, as he had across Europe and the English-speaking world and as he hoped to in the major markets of Asia. Murdoch was baffled in his 2000 attempt to take over GM's DirecTV when its smaller rival, Charles Ergen's EchoStar, agreed to combine with DirecTV (the deal was subsequently rejected by the Federal Communications Commission, even though it would create a company smaller than Comcast–AT&T). Despite the failure of that attempt, Murdoch still threatened what cable monopolists feared most: a sustained, coordinated attack on local cable franchises by cut-rate competition, led by his Fox news, sports, and entertainment networks, which aggressively courted friends in Republican Washington, D.C., and other high places. To the cable fraternity, which could fight and make up like drunken lovers, Murdoch had long ago slipped beyond good company, through his incessant attacks—in Congressional testimony and press interviews—on cable pricing and programming practices, and even on cable personalities like Ted Turner, whose sanity Murdoch's *New York Post* questioned from time to time.

Brian Roberts said he'd cut costs by ending staff at TCI's headquarters, from negotiating cheaper deals with programmers, and from creating unspecified "operating efficiencies." "This is the deal of a lifetime," Roberts told investors. "Think about the enviable position that this company will have in content and in national advertising. . . . For years we are going to have a platform for both communications and for entertainment that is unrivaled."[17]

Plus a debt load of nearly $40 billion. Roberts acknowledged the gamble: "In doing this transaction it was clear that you had to make two basic bets. One was that we would be able to improve the cash flow. . . . Two was that the non-strategic [stock] holdings of AT&T could be monetized," or sold off. "We think we can do that."

Smokescreen

Comcast's July offer had thrown a scare into consumer advocates who'd been battling the company over its exclusive right to sell Internet service on its cables. The merger "would have harmful consequences for the future of the Internet," warned the Center for Digital Democracy in Washington. The center's prime concern: Comcast's successful resistance to being classed as a "common carrier" instead of a private, members-only system. Common carriers like phone companies and electrical power distributors are generally open to anyone who wants to plug in and rent them. Cable argued that it was different. Comcast said that it might well allow certain qualified competitors to use its network, but it shouldn't be compelled to. Though the FCC in past years had forced cable carriers to carry all the broadcast channels in their neighborhoods, under Chairman Michael Powell the commission was not inclined to force Comcast and its peers to extend this into, for example, carrying competing Internet services. It was unfair, howled the phone companies, and the would-be cable competitors, and the consumer groups. And wouldn't letting Comcast buy AT&T make things worse?

"This is a giant step toward Comcast controlling the eyeballs and the wallets of millions of consumers and citizens who will not have the ability to turn elsewhere for their information," the center added in its critique of the deal. "It is a deal that only a Comcast executive could love. They get a cable monopoly to help build a television and high-speed Internet empire. Everyone else gets less choice, higher cable rates, and an Internet that resembles a giant advertisement." Asserted center director Jeff Chester, "Comcast wants to become an unregulated digital toll booth, and it will use its dominant monopoly status to extract new fees from competitors and consumers alike."[18]

Comcast also drew sniping from consumer activists in Congress. Sen. Barbara Boxer, D-Calif., blasted Comcast's "monopolistic" practices in a letter to the SEC. Senators Herb Kohl, D-Wis., and Mike DeWine, R-Ohio, worried that Comcast and other big cable companies could "pose a threat to innovation and diversity" and "hamper efforts to foster competition." Comcast acknowledged and dismissed their concerns, and the Senate took no action to block the deal.[19]

Likewise the credit-rating agencies—Standard & Poor's, Moody's and Fitch's—wrung their hands and, in some cases, cut Comcast's ratings but kept it above junk bond status, effectively endorsing the deal for the banks that lent Comcast $30 billion to make the AT&T deal possible.

On Wall Street, the biggest firms, including Citigroup, Merrill Lynch, and Morgan Stanley, were hired by Comcast and AT&T to evaluate the deal, and were paid tens of millions of dollars in fees for their favorable opinions, not counting the millions more they stood to make by trading Comcast bonds or lending the company money to finance its debts. Even after it was done, the merger attracted sneers from some less interested parties. "The deal stinks for everyone but Brian Roberts," groused Andy Kessler, a former Silicon Valley telecom stock analyst and author of *Wall Street Meat*, a highly readable first-person account of the Internet securities bubble. In columns for the *Wall Street Journal*, Kessler slammed the AT&T deal and broadened his attack to include all government-guaranteed cable monopolies, ridiculed Roberts's vision of "distribution control" (which he called vulnerable to wireless, fiber, and satellite competition), and called Comcast's profit projections a "smokescreen" to hide the real goal of all cable mergers: "Cable is an entire industry structured around tax avoidance. . . . They are nearing the end of a depreciation cycle, and either they [buy AT&T] now or they sell. But of course they can't sell because then Brian Roberts would be out of a job."[20] And it wasn't just gadflies, short-sellers, or hedge-fund managers who looked down their noses. "These cable companies are like giant tax shelters," said a senior executive at one of the nation's largest pension investment firms. "It's a huge black box. None of them have reported earnings. None of them have paid any taxes." But, he added, it seemed Comcast wasn't likely to turn out to be another Enron: "At least in this black box, there is a real business."

Still another critique came from the consultant hired to review the deal by Minneapolis and other towns in three states. The firm of Creighton, Bradley & Guzzetta and its accountant, Ashpaugh & Sculco, expressed alarm at what it called the lack of solid financial information from Comcast concerning the likely results of the deal. Most of the towns readily approved the deal anyway; municipalities had by now lost most of their cable regulatory powers, with the changes coming primarily during the years of Republican dominance of the FCC. But the challenge was reported in the trade and regional press, and Comcast took the accusations seriously enough to send the Creighton firm a detailed letter ridiculing the firm's assault on what the company called its "common sense proposition" to buy AT&T. Comcast Executive Vice President David L. Cohen was more direct: He accused the firm of going negative on purpose, to give its client towns leverage in franchise negotiations with Comcast, a charge Creighton denied.[21]

In the end, once Comcast decided to do the deal, the only opinions that mattered were AT&T's acceptance, the credit agencies' endorsement, and the FCC's approval. Writing for the majority, Chairman Powell approved Comcast's plan in November 2002, nearly a year after the deal was struck. He imposed one condition: the isolation and eventual sale of Comcast's investment in Time Warner Entertainment, which could, Powell noted, create a conflict of interest with Time Warner's America Online division as Comcast built up its own rival Internet service. The outspoken Democrat on the commission, Michael Copps, objected to the combined companies' market power in a letter three times longer than Powell's. But the deal was all over—except for the job of integrating two enormous companies.

"We will go from a regional cable company to being a premiere provider of entertainment and communications services into people's homes," asserted Brian Roberts to *Business Week*. "If we paid a full and fair premium for the company, why should we give up control? It's been good so far for shareholders," since Comcast went public in 1972.

"I don't get to relish this deal. I have to deliver."

What They Paid For

Brian Roberts had agreed to pay nearly $5,000 for every AT&T subscriber. If the company had zero costs and all its customers bought premium services, it would still take six years to make that money back. But Roberts agreed to pay in stock, in a market that was going down, down, down. By the time the deal actually closed, Comcast stock had lost a third of its value, and that made the purchase that much cheaper for Comcast.

"No one will have more control over which TV channels, Internet services, and movies are piped into U.S. homes—not News Corporation's Rupert Murdock, not Sumner Redstone at Viacom Incorporated," *Business Week* wrote. "Right from the start, [Brian] Roberts will have the clout to do what cable executives have wanted to do for years: dictate what shows will reach a mass audience and at what price. . . . For years, content-driven companies such as Walt Disney, News Corp. and Viacom have had the upper hand. That's about to change."[22]

Brian Roberts told where he was headed to anyone who pointed a tape recorder in his general direction that year: "Why only be a cable company? I really think we are at the cusp of a new golden age of talking television and computers and all these connections to a whole new level." That summer, Brian was negotiating to buy controlling interests in the Golf Channel and *Outdoor Life* to supplement its partnerships with Time Warner (QVC) and Disney (E! The Entertainment Channel).)[23] *Business Week* went on to speculate that Comcast was sure to buy a programming company next. It asked, Why not Disney?[24]

COMCAST'S CRITICS

No Sense

The history of cable television can be read as a series of political, technology, and marketing campaigns to keep prices high for a service that could otherwise be delivered at a more competitive cost. It's a story as old as business and politics: Railroads, oil companies, grain processors, and drug makers have done as much and more, when they could get away with it.

At the quarterly cluster of investor conferences where he makes his most eloquent defense of what he does for a living, Brian L. Roberts spends much of his time preaching against price wars and defending the company's resistance to the laws of supply and demand. "We can always cut price. That's the easy, quick thing to do. [But] I don't think it's our first choice," Roberts has said. Comcast will emphasize speed, or availability, or pair its service with cable-based telephone systems, or some other combination to avoid charging consumers less: "People are waiting for the big price cut and I hope it doesn't have to happen." As long as most customers are willing to pay more, he's argued, why shouldn't they?[1]

Like other cable companies, Comcast prefers to deflect blame for its fat yearly price hikes to others—to the people in Hollywood who produce television programs, whose fees are Comcast's largest single expense, costing more than its entire payroll. "It is a competitive business, and there are real cost pressures on the cable companies," according to Roberts.[2]

"It doesn't seem to me to make any sense at all, when the basic inflation rate is one or two percent, for a programmer to come in the door and demand seven, ten, fifteen, twenty percent," said Steve Burke, head of Comcast's cable division, at a summer investors' conference in 2003. "We spend more on programming costs than we do on people, and we have 55,000 employees. . . . That's a dangerous thing."[3]

Just how dangerous? According to Comcast's financial report for 2003, published five months later, Comcast wasn't paying increases anywhere close to Burke's range. According to Comcast, its average monthly bill—including the many customers who signed up for Internet and other special services, where there are no programming costs— was up 8 percent for the year, to around $65. The average basic cable bill was up 5 percent, to $45. But Comcast programming expenses had risen just 2.3 percent—less than half its cable increase and a third of its typical customer bill.[4]

That was an unusual year, in which Comcast lawyers had aggressively jawed down programming cost hikes for the former AT&T cable units it had absorbed. In the future, Roberts estimated, these costs might rise, perhaps 10 percent a year. But other Comcast costs, he promised, would go down considerably, now that most of AT&T's old system had been upgraded. Yet he still wasn't projecting rate relief. To the contrary: "I think the industry needs always be vigilant on prices."[5]

Monopoly in Motion

How can Comcast charge more if its key operating costs are going down? Consumer advocates have an easy explanation: Despite broadcast and satellite, dial-up cable and wireless, cable companies still face no significant competition for their core customers, the typical television-watching American household. "The cable TV industry is doing exactly what should be expected from a monopolist whose hand had just been strengthened" by giant mergers, wrote *Consumers Union* in its 2003 review of cable rates.[6]

Where were customers' dollars going? To set up new and still more profitable services like cable Internet, and to buy other cable companies, *Consumers Union* continued: "While the cable industry has certainly increased capital expenditures to upgrade its plants, it has actually sunk a lot more capital into another activity—mergers and acquisitions. It is the outrageous prices that have been paid to buy each other out and consolidate the industry that [are] helping to drive the rate increases." From 1998 through 2002, cable companies spent a total of less than $20 billion a year improving their systems, and over $60 billion each year (in cash, debt, and stock) buying other cable companies.[7] *Consumers Union* called on Congress "to treat cable monopolies the same way they treat telephone monopolies," or face a future of "fewer choices, rising prices and reduced customer service."

Who's Paying for This Microphone?

The Internet "is the closest thing ever invented to a true 'free market' of ideas," and it's in danger from Comcast and its fellow cable giants, the American Civil Liberties Union declared in a 2004 report ominously titled, "No Competition: How Monopoly Control of the Broadband Internet Threatens Free Speech.

What exactly was the threat? Cable companies, unlike phone systems and highways, were exempt from the "common carrier" designation that forced other utilities to serve any potential customer. Cable also provided the best, fastest, and, increasingly, the most popular form of connection to the Internet. To seize such opportunities, "the cable industry is trying to extend its old business model to the broadband Internet, leaving consumers with little choice about where to get their high-speed access—it is their cable provider or no one at all." The ACLU painted a dark picture of cable companies both tracking users' Web browsing and blocking access for commercial, political, or competitive reasons, "like the phone company being allowed to own restaurants and then providing good service [to] Domino's [but] frequent busy signals, disconnects and static for those calling Pizza Hut." In short, cable companies "have both the financial incentive and the technological capability to interfere with the Internet as a free and neutral medium for the exchange of information. . . . What will hold them back?"

Comcast and its cable allies have fought hard to prevent its cable service from being reclassified like a phone company's—so far successfully. Warning that anything can happen "in an election year," Brian Roberts praised the way the "government does sort of understand that it's a different business than it was ten years ago," and is "more likely to have less regulation on the new technologies," in order to "give you incentives to invest in . . . less regulated" businesses like cable internet and video-on-demand.[8] In February 2004, the cable trade group, the National Cable and Telecommunications Association, wrote to the FCC arguing that since "cable operators have no intention of blocking access to content," no law is necessary.

What of the ACLU's charges that private control meant limits on "free speech"? Qwest Communications chief executive Dick Notebaert told Roberts, in a letter, that Comcast's refusal to run Qwest's ads for competing Internet service "smacks of censorship. . . . By denying us access, you are blocking a customer's ability to learn about all the choices in the marketplace and to make an informed decision." Comcast responded that it accepted Qwest ads—just not, apparently, its Internet advertising.[9]

Meanwhile, a New Jersey–based antiwar group complained that Comcast had accepted its ads, then pulled them at the last minute, in the build-up to the invasion of Iraq. Peace Now of Princeton spent $5,000 for CNN airtime that would have showed citizens opposed to the war. The group said smaller cable companies had run the spots in Midwestern markets without protest, and complained of Comcast's "infringement of our First Amendment rights" around the nation's capital and other big Comcast markets. "We must decline to run any spot that fails to substantiate certain claims or charges. In our view, this spot raises such questions," Comcast spokesman Mitchell Schmale responded.[10]

On the Cheap

Cable didn't have to cost what Comcast charged. Just outside "Comcast country," customers in the little Pennsylvania college burg of Kutztown, in 2003, paid just $30 a month for 88 channels of "expanded basic" service. Put another way, giant Comcast was charging $6.44 per channel per year, while little Kutztown charged just $3.60 per channel. Generally, big cable companies like Comcast charge the highest prices. Regional cable companies like the Walson family's old Service Electric, which serves customers in the small cities and resort and former mining towns north of Kutztown, charge less—around $35 a month. And rates are lowest in towns like Kutztown, with nonprofit or municipal services; or Troy, Alabama, with multiple cable providers; or the Massachusetts and New Jersey towns that allowed fiber-optic cable operators like RCN to compete with Comcast. (And these services aren't primitive; even little Kutztown offers a long menu of cheap data and high-speed Internet services.)

But just as few cities still have their own gas or electric companies, extensive lobbying by Comcast and other cable companies has prevented towns from offering their own cable, or even permitting competition. In three Chicago suburbs during the fall of 2002, cable lobbying helped convince a majority of voters to reject the town fathers' plans to build their own high-speed Internet network: Taxpayers weren't willing to trade the up-front cost for the possibility of cheaper or more flexible long-term service. Managers of the three Fox River Valley towns found themselves campaigning in vain against a cable-backed "Vote No" campaign. Comcast kept its monopoly.

You didn't need municipal socialism to show cable costs could be made to drop. Competition would do the job, in the 2 percent of markets that provided any, the Federal Communications Commission argued in yearly reports it traditionally releases during the Christmas holidays when no one's paying much attention. Microsoft founder Paul Allen's RCN Corporation, backed by $5 billion in boom-era venture capital, was providing much of the proof. In Boston, "in response to RCN's entry," the dominant Cablevision franchise "moderated its regional rate increase" and "agreed to improve its commitment to public and educational channels," the FCC reported. In nearby Somerville, Massachusetts, Time Warner "announced a rate freeze for only that area" when RCN moved in. In northern Virginia, Cox "announced an upgrade" when RCN won the right to compete. Comcast's customers also got a break wherever RCN came to town. In the Washington, D.C., area, as a result of RCN's arrival, Comcast "reduced a previously proposed rate increase." When RCN managed to gain a foothold in Philadelphia's Delaware County suburbs, Comcast "began to offer 'rate locks' and service improvements in the towns where it faced competition," according to the FCC report. (In RCN's case that wasn't enough. The company spent far more time trying to open new markets, and less laying fiber, than it had expected. In 2004, RCN filed for bankruptcy reorganization, though its systems remained in business.)

Consumer Outrage

Just how bad is Comcast's customer service?

In 2003, the 100,000-member National Quality Research Center ranked Comcast below even the despised Internal Revenue Service, and tied the company with rival Charter Communications for the worst rating of the airlines, utilities, and government services evaluated. "That doesn't mean that people enjoy paying taxes more than they do watching cable TV. But in the context of what these organizations do, the IRS offers more satisfactory assistance and, unlike the cable companies, the IRS has no 'price' to increase," explained the center's director, Prof. Claes Fornell of the University of Michigan.

J. D. Power and Associates, best known for rating cars, reached a similar conclusion. "Digital and analog cable subscribers . . . dramatically trail satellite subscribers in overall customer satisfaction," thanks in part to cable rates, which had grown five times faster than satellite charges over the previous five years. Even among cable providers, Comcast ranked below average, tied with Charter and Cablevision and just above Adelphia, whose founders had recently been arrested on corporate fraud charges. (By contrast, Cox Communications and RCN ranked above average, though still below the two satellite services, DirecTV and DISH.)[11]

Around the same time, the Wall Street brokerage S. G. Cowen & Company ran its own customer service survey of 2,700 households and found a majority of satellite TV subscribers felt "very satisfied" with their providers, DISH (formerly EchoStar) and DirectTV. Only one in four cable customers felt the same way. Similarly, six in ten satellite customers got "good value" for their monthly payments; only one in three cable customers felt that way.[12]

Even *PC Magazine* ranked Comcast at the bottom: ninth out of nine in a survey of Internet services.[13]

"Bad for Business?"

Comcast's purchase of AT&T Broadband made it the nation's largest "porn-promoting corporation," declared the ecumenical Religious Alliance Against Pornography on the eve of the merger. Comcast, like AT&T, is responsible for "the mainstreaming of pornography," the Roman Catholic Archbishop of Baltimore said.[14] The cardinal and his allies blamed the companies for "taking it out of back-alley smut shops and channeling it directly into people's living rooms." The group also took on DirecTV, complaining that the service (then controlled by General Motors, subsequently by Rupert Murdoch's News Corporation) "sells more videos than Larry Flynt," publisher of raunchy *Hustler* magazine.

"Despite the profitability of this part of the cable business, we hope to show them that pornography is bad for business and for their corporate reputations," said coalition president Rick Schatz. He called on members to write to Brian Roberts and ask him to reconsider carrying "soft- and hard-core pornography" through the Playboy Channel and

Hot Network. The coalition had tried to work through polite channels, but said it made little headway, "largely because the multibillion-dollar porn industry is so profitable."

Ralph Roberts had once had qualms about carrying dirty movies. But Comcast had long ago gotten over it. In that respect, the company was giving many Americans what they wanted.

BEYOND DISNEY

Who Wants a Price War?

It wasn't enough for Brian Roberts just to win AT&T. His company was barely done digesting its larger rival, when Roberts went gunning for another huge trophy, Walt Disney & Company. Like AT&T, Disney was one of the great names in American business, the closest thing there has ever been to a beloved corporation, with its copyright grip on mass-market childhood folklore and family escape; it was part of the Dow-Jones Industrial Average, its framed stock certificates embellished with cartoon characters hung in many an ambitious parent's nursery, an American emblem of fortune, fantasy, and wishing upon a star. After months of back-door attempts to outflank Disney chairman Michael Eisner and force his surrender, Roberts went public with his hostile offer in February 2004. TV and newspapers called the bid a shocker. But for people who had been paying attention to Brian Roberts all along, it was hardly a surprise: Comcast hadn't even taken full title to AT&T's cable monopolies in the fall of 2002 when the business press, at least, was already speculating "about Roberts buying Disney some day."[1]

The Comcast–AT&T deal was the high point, and nearly the end, of the long string of mergers that had created a handful of giant cable companies, with Comcast the biggest of all. But five months after that deal closed, Rupert Murdoch, owner of News Corporation, provoked another round of high-stakes deal making by finally buying control of DirecTV, the biggest U.S. satellite television supplier, for $7 billion. That purchase, though just a fraction of what Comcast had paid for AT&T, relit the cable industry's fear of Murdoch's multinational media conglomerate and its threat of price-cutting, consumer-friendly competition. Murdoch, with his record of deep-pocketed spending on programs and products, and aggressive price cuts, threatened finally to tilt the television business away from the lucrative cable monopolies by offering a competitive and attractive alternative.

It had already worked in Britain. Murdoch had blown Comcast and other cable investors off the island in the 1990s with his British Sky Broadcasting (BskyB). With his publishing fortune behind it, satellite had triumphed over British cable because it was cheaper and offered more programs. Of course, Murdoch wanted to duplicate his success in the world's richest market, the United States. But Murdoch failed in his first attempts. His initial efforts to buy a U.S. satellite system ran into political and legal problems when the cable operators, who considered him a genuine threat, actively opposed him. The cable companies used the delay to improve their networks and add Internet and video services, which they priced in such a way as to punish anyone who tried to switch basic TV service to satellite. Many of cable's improvements in those years—its investment in fiber-optic lines and digital recording gadgets and Internet service—were made with an eye toward stopping Murdoch's eventual assault. Murdoch, for his part, beefed up his news content and added what his critics claimed was something new in mainstream U.S. media: an obvious bias toward, not only political conservatives, but particularly the party in power, the Republicans, which helped the nation's powerful see Murdoch in a new light. Now, with the invader at hand, Roberts and his peers took to mentioning "Rupert," in their public pronouncements, among their biggest challenges.

Everything about Murdoch seemed extra: his three wives; the way he set his two eldest sons, Lachlan and James, into executive positions, over British shareholders' howling, impotent protests (it is a measure either of U.S. shareholders' weakness or of Ralph Roberts's smooth ways that Brian's promotion never provoked similar outrage); Fox-brand U.S. movies, television studios, and local sports; the diversity of his aggressively partisan U.S. news commentators, bare-chested British newspaper pin-ups, and manic cartoon shows; thirty-five U.S. TV stations; 100-plus newspapers around the globe, anchored by the *Times* of London and the loud tabloid *New York Post*; Australian, British, European, Chinese, and Indian TV and publishing enterprises; and tax shelters around the world.

Total sales for News Corporation weren't too much more than Comcast's. But if Murdoch wanted to mount a real competitive challenge to cable, his pockets were deep enough. Roberts fought back by promising a menu of services satellite couldn't yet offer: high-speed Internet, video-on-demand, and local sports that he could withhold from DirecTV in markets like Philadelphia. Feeling pressure to seem agreeable so he could get his DirecTV deal approved, Murdoch even went so far as to deny he wanted a price war with cable—though he couldn't, or didn't bother, to tell this to his U.S. lieutenants: "We would love a price war, to have a bunch of people saying we have to give broadband away," said his top DirecTV executive, Chase Carey, on the eve of the DirecTV deal.[2]

If Murdoch, the aggressive competitor, was cagey about his intentions toward Roberts and cable, Brian Roberts, the monopolist, was frank about the Murdoch threat. "He has broadcast, news, sports, movies, cable channels, and now [TV] distribution. You'd have to be crazy to not take all that seriously," he said in response to the DirecTV deal. "That's going to cause a lot of people to reassess their business."[3]

And so, under President George W. Bush's Federal Communications Commission chairman, Michael Powell, Fox–DirecTV followed Comcast–AT&T onto the "Approved" list of big media mergers. It seemed a cynical contradiction that, a year before kissing Murdoch, the FCC had forbidden one of the last really independent figures in American television, Charles Ergen and his EchoStar (later DISH Network) satellite system, from buying DirecTV. Powell's logic: Combining DirecTV and DISH, the two main satellite systems, would have ended satellite competition, while selling one to Murdoch would bring more investment capital into the industry. And capital, the Bush FCC maintained, was always good for competition.

The FCC did set some conditions that hamstrung Murdoch in his fight against Comcast and the other cable companies: Murdoch promised, for four years, not to bar any cable system from buying Fox programs, and to accept arbitration in case of disputes over cable service until 2009.[4] And Murdoch got a key exemption: The FCC agreed not to force him to share his pro football games, reasoning that he didn't own the National Football League, but had paid a fair price for controlling its weekend games. And that, of course, tilted the field toward cable: Roberts was under no corresponding obligation to start letting satellite TV competitors show pro sports that his cable company controlled.

What Murdoch Didn't Have

While Rupert Murdoch was busy firing DirecTV executives and putting his own in place, Brian Roberts and Steve Burke were hard at work forcing

AT&T into the Comcast mold. After firing more than 5,000 workers, they held mass rallies at Comcast offices all over the U.S., opened new customer-service centers, and exhorted contractors to work harder. Complaints poured in, as they always do when big companies merge: local newspapers delighted in showcasing such familiar cable problems as the Washington widow repeatedly billed for her dead husband's service, or the New England affiliate that inadvertently switched porn for family shows.

But on Wall Street, among the bankers and investors whose goodwill Roberts felt he needed at least as much as his customers', Comcast impressed even some of the skeptical bond buyers—who care not about a company's hyped-up prospects, the way stock investors do, but about cold, hard cash, sales, profits, debts. The boom in cable Internet helped replenish the bottom line, as Comcast grew quickly to more than 5 million Internet customers, trailing only America Online and Microsoft. Comcast also stepped up efforts to sell national cable advertising. And the company bargained down programmers like Viacom and ESPN, which had collected double-digit fee hikes in previous years, often dropping rates AT&T had promised to the lower levels Comcast was accustomed to paying. Comcast also cut its reported costs by declaring that repairs to the AT&T system would cost less than originally estimated. And, most important of all, it sold QVC to John Malone's Liberty Media, collecting nearly $8 billion, enough to wipe out the deficit and report a record profit, at least for 2003.

To make sure its message got out, Comcast held a series of increasingly triumphant conferences with professional investors. "This is a historic day for us," Brian Roberts said in May 2003 at one such gathering in New York. "The early results are very, very promising. . . . The cable model is truly working. . . . My number one goal is to maintain the entrepreneurial spirit that has always defined Comcast. . . . It's only going to get better."[5]

Comcast cable chief Steve Burke, a former Disney executive, picked up the thread in San Francisco four months later. With video-on-demand, which allows digital cable customers to see shows long after they have been broadcast, cable finally had something satellite couldn't match, Burke said. (He didn't mention another something its competitors lacked: Comcast's local sports monopolies.)[6]

Burke did more than refine the well-honed company line. He said Comcast faced a dilemma over how to handle Murdoch: "The big debate within our company is, do we want to compete with [price-cutting DISH Network's] Charlie Ergen, or with [program-heavy News Corporation's] Rupert Murdoch?" (Who, at Comcast, could possibly have challenged Roberts's well-known antipathy to cutting the prices customers pay?

"Must have been someone in marketing," said a house skeptic at Comcast headquarters. "They really like to get those subscribers up.") Comcast, Burke said, had used the time gained by the frustration of Murdoch's early U.S. expansion plans wisely: improving the cable in old AT&T systems, and marketing high-speed Internet and experimental video-on-demand, which he hoped would prove addictive, to customers who would not want to leave. Burke continued, "We're in a position to be a lot more competitive" than the cable companies Murdoch had beaten abroad. For example, video-on-demand "allows us to go to people who have a dish and ask them, 'Why don't you go back to cable?'" He added, "The gloves are off."

Yet Comcast clearly preferred to avoid all-out competition. Better to link the company's future, or at least its claims, to some technology or tie-in that would-be competitors couldn't meet. Michael Armstrong had proved you couldn't simply buy your way to market power. He had bought everything he wanted; yet he lost control of AT&T quickly when he couldn't produce new customers or higher profits—or even hold what he had. In the mainstream cable tradition (software companies do it, too), Comcast would keep investors dazzled with the promise of something new. It was worthwhile, at least, to keep people guessing: Would any of the new products go beyond mere gadgetry and increase viewers' willingness to keep paying more to remain Comcast customers? Since Philadelphia had been wired and the nation more or less fully, if not competitively, cabled in the mid-1980s, cable personalities had been selling Wall Street one vision after another in an attempt to maintain investors' interest: Five hundred channels! Interactive cable! Cellular synergy! High-definition! Digital! Internet! Video-on-demand! What would Comcast roll out next?

Beyond Reality

"It was a fantasy year," Brian Roberts told the media-company investors who followed his quarterly traveling show to the sprawling Biltmore Resort and Spa in Phoenix at the beginning of 2004.[7]

Yes, the Biltmore, named after New York's railroad millionaire Vanderbilts, just like the old Manhattan hotel where Brian's immigrant grandfather ran the pharmacy flagship of his own little fleet of hopeful investments so long ago. But that was nothing to compare with Brian's role at the center of his entourage of executives, investors, and analysts. Here he was, spending long busy days at this warm winter paradise, surrounded by people who wished him the success that would make them all richer.

The national American narrative is the bragging story of one's own prosperity, achieved by overcoming great obstacles. It is not unusual to magnify those obstacles so as to make one's achievement all the more remarkable. Comcast makes much of its humble beginnings in a small Southern town with 1,200 subscribers; that achievement falls into context when one realizes the considerable Philadelphia capital, experience, family resources, and professional network mobilized by and behind Ralph Roberts's energy and smarts. Now Brian was emphasizing that Comcast had started the previous year with everything going against it: a debt of more than $30 billion; fee disputes and courthouse battles with Liberty Media and other programmers; and customers fleeing the run-down cable systems he'd just bought from AT&T. Of course, those obstacles were of Comcast's own making, as Roberts acknowledged: "We had a feeling, 'Oh my gosh, what did I just do?'" But now Roberts had the numbers in hand to proclaim his mission accomplished—sales, customers, and profits all at or above target, thanks in no small part to aggressive cost cutting and the sales of some important (but mostly noncable) provinces of the Comcast empire. By those measures, Roberts could well proclaim the victory of his previous audacity, before an audience he knew was in his pocket: Professional investors had just voted him America's best chief executive, in *Institutional Investor* magazine's yearly poll of pension plan and mutual fund managers.[8] Comcast's co–chief financial officers—quick, cutting Larry Smith and polished John Alchin—had scored similar honors, apparently laying to rest any old doubts about the company's 1,000 subsidiaries, complex tax shelters, and phone-book-sized financial reports.

Yes, quite a year. Comcast had proved its serious intent by selling a string of assets to raise cash; it had shrunk its debt by a third, and warded off the unending efforts by cable's enemies to get Washington regulators to put their heavy boots on cable rights and cable rates. The auditors back in Philadelphia were busy preparing the annual report that would show record profits, thanks largely to the sale of QVC but also to tight cost controls and surging fees. And Comcast could claim it was approaching its proudest target: The company said it was keeping close to 40 cents of every dollar in cable fees, and half of every Internet dollar. That wasn't real profit, of course. That was counting the way Comcast likes to count, which is without counting taxes, interest, depreciation, and other annoying inescapables.

So the shining numbers depended largely on the cable industry's old accounting magic. Meanwhile, the all-too-real expenses of upgrading cable and keeping all those trucks off the road and paying 60,000 regular paychecks and paying off the 5,000 ex-AT&T people the merged company had dropped were, in total, actually greater than Comcast's billings brought in. But not for long, Roberts insisted: Voice-over Internet protocol, which makes your TV line a phone, and video-on-demand, and all those other new services on the way, would surely swell future Comcast profit reports, even if, as killjoy analysts kept warning, cable sign-ups were flat and new Internet sign-ups were finally starting to slow down. Though high-speed cable Internet had become a billion-dollar item in Comcast's sales book, analysts warned that so many people signed up in 2003 that cable companies would be unable to repeat that year's growth. The companies would likely need something else to keep investors interested. One thoughtful cable analyst made that point in a closely written twenty-nine-page report—issued the day before Roberts bid on Disney.[9]

But Roberts wasn't talking to the Arizona crowd about Disney. He was still promising higher profits from existing, cable-based businesses: "I'm cautiously hopeful we have blue skies ahead." A little more emphatically: "We are at the promised land of free cash flow." Yes, media executives do talk like that.

Of course, Roberts quickly added, there were clouds on this blue horizon. Satellites, even. "Rupert," again. But Comcast, he assured them, was up to the challenge.

All right, said the man from Citigroup Smith Barney, the world's largest bank and Roberts's host; but what about customer service?

"We have gotten medium grades in some of our reports that we get from third parties," Brian acknowledged. "The defensive part of us says 'hey, we're doing the best we can,'" he couldn't help remarking. But in "the real world," he admitted, "we have to get going."

It is, he declared, "time to tell our customers and consumers that this is not the same old Comcast."

"A Logical Next Step"

One month later, Brian L. Roberts stood before his investors again, and marked 2004 as an even bigger fantasy year: He announced his plan to buy Walt Disney Company, over chief executive Michael Eisner's objections. The Comcast bid for Disney went down a little like a South American novel—a deal foretold in advance by the national business press and the trade papers and Brian Roberts's own foreshadowings, yet treated like a shocker in the day's news. *Business Week* wasn't the only publication to preach a Comcast–Disney alliance before the fact. The *Daily Deal* listed among its "Predictions for 2004" a Comcast all-stock bid; it even supplied a happy ending, in the form of a sweetened offer and Eisner's voluntary admission that "suddenly, it's time to call it a day."[10] None of which stopped the Associated Press, *Boston Globe*, *Los Angeles Times*, *Times* of London, and *Wall Street Journal* from declaring the offer a "surprise" when it actually went down.[11]

"I don't think this is a huge surprise," countered Brian Roberts. "It's a logical next step."[12]

In fact, the offer's hostile nature did recall certain other Roberts attempts over the years: the abortive back-door attempt to join Microsoft in quietly muscling into John Malone's TCI, after the more successful efforts to take over the much smaller Jones and Lenfest companies by similarly devious maneuvering before their founding owners knew a bid was even on the table. Making the offer public was also, quite literally, a logical next step; Roberts had indeed been stalking Disney for months, using his old network of Wall Street intermediaries to sound out Disney's "independent" directors, hoping they would prove as disloyal to Eisner as he had once hoped Magness's survivors would be to Malone. After weeks of probing Disney's board by proxy, Roberts came to believe its directors could, indeed, overthrow their current leader in favor of Roberts and his team, plus Comcast's cash; the board might see in the cable company a fresher and more promising alternative to Eisner, who had alienated the Disney family and whose recent record compassed mediocre movies, weak theme-park sales, and the weakest major TV network (ABC). Yet when Roberts finally called Eisner to make a deal, Roberts appeared to act, not from cold, calculated, irrefutable logic, but with a certain gut-level impetuosity: The onetime cell-phone titan walked out of a meeting of his aides in a Philadelphia hotel and called Eisner from a pay phone to make his offer. But Eisner was ready for him: The Disney chief had prolonged a highly controversial and very highly paid career far beyond the typical CEO term precisely because he was very good at shooting back.

You're Buying What?

Roberts tried to put his offer in the perspective of all the Comcast deals that had gone before: "It looks like the Disney board members are considering succession, and this was an opportunity to act upon." He added, "I think being at the right place at the right time has been one of our hallmarks."[13] In 2001 and 2002, Brian Roberts had presented the

AT&T deal as a once-in-a-lifetime chance to buy Comcast into the big leagues. AT&T had spurned him at first, but after many quiet meetings the companies had come to terms. Now, here he was again, trying to sell the Disney deal—another $66 billion tie-up, or nearly so—on its merits. "This," he declared "is an incredibly compelling combination." "Compelling," and "brilliant," gushed the analyst from Merrill Lynch, Jessica Reif Cohen, whose firm nevertheless was passed over for the lucrative investment banking business it had enjoyed with AT&T. (And it was lucrative: Comcast had spent $222 million on the AT&T deal. And that was just for investment bankers. The lawyers were extra.)[14]

Comcast offered 78 percent of a Comcast share for every Disney share. Comcast promptly sank, and Disney rose—proof investors expected Comcast would increase its bid, maybe even to a whole share, maybe with cash, despite Roberts's denials. Michael Eisner, Roberts complained, had rejected the bid out of hand. So Comcast had taken its case to the public, most especially to Disney shareholders, who had grown restive under Eisner's twenty-year hold on the company.

Eisner had come to power in an earlier hostile bid from another Philadelphia company, Saul P. Steinberg's Reliance Insurance,[15] in one of the classic "greenmail" takeover attempts of the 1980s. Steinberg had accepted $60 million to go away (a judge made him give back half), but the shock had made Roy Disney, Walt's nephew, and his allies understand the need for more aggressive management; that led them to Eisner, to a decade of rapid expansion and growing profits, but later to high costs, weak growth, declining investment, and a new movement by Roy and his allies to unseat Eisner, after all those years. Comcast, in its old Ralph bargain-hunting mode but on Brian's greater scale, hoped to capitalize on that discontent, and on Disney's weak share price.

What did a cable company know about Disney? "We are very fortunate" that Comcast Cable President Steve Burke "worked at The Walt Disney Company for 12 years," at AT&T, Euro Disney, and the retail unit, Roberts said. Indeed, Burke's father had helped run ABC years before; the younger Burke had grown up in the business, like his current boss. Roberts added, "We are very respectful and mindful of the incredible brand and the institution that is the Walt Disney Company." With Comcast, he said, the companies would be "a leader in all four spaces of family programming, films, news and sports, nationally and locally, with an unmatched technology platform in a very competitive business." Burke even offered a breakdown on how Comcast could save a billion dollars by cutting back-office jobs, and told how it could boost the performance of Disney's ABC network to the level of its more profitable competitors. And, Burke said, in an appeal to any Disney executive who had ever felt crowded by the acerbic, micromanaging Eisner, Comcast knew how to leave a business to the people who knew how to run it—as it had, more or less, with QVC, and cellular, and SportsNet.[16]

The offer got plenty of headlines, but it wasn't enough money for Disney's board. So the waiting game began.

"Highly Underwhelming"

Eisner told the story on cable television, on the Cable News Network's chatty Larry King Show.

"Roberts called you, right?"

"He called. I told him I thought the board felt we were on the right strategic course. I didn't get much else other than that out. And we got an offer. And the board...determined it was highly underwhelming."

What about a better offer?

"We will accept it," Eisner said. But, "it has to be pretty spectacular," to include Disney theme parks and Disney studios and ESPN.

And, King asked, wasn't ESPN, after all, the magnet drawing Comcast to its bold, or desperate, bid: the billions Disney extracted from cable for carrying its programs each year?

Spoon-fed such agreeable questions, Eisner crowed over Disney's newly concluded cable deal with the Cox cable network: a big first-year increase, followed by eight more years of modest growth.

"Do you expect the same from Comcast?"

"Well, I would like more from Comcast, actually."

The studio audience laughed.[17]

Wishful Thinking

What was Brian Roberts thinking? The Comcast chief was convinced he could drive a wedge between abrasive, tenacious, increasingly unpopular Disney CEO Michael Eisner and Eisner's own handpicked directors, who were led by former U.S. Sen. George Mitchell, best known to the public for his role as a peacemaker in Northern Ireland, but plenty prominent in worldly Washington circles after his retirement from Congress. Mitchell had known Roberts well enough to invite him to address the Economic Club of Washington in early 2001, praising him, his father, and Comcast in a ringing introduction before the political–financial elite of the nation's capital.[18] Now Roberts, through his advisors, came to believe that Mitchell was open to a deal behind his boss's back.[19] Either Roberts and his helpers misread Mitchell, or Roberts's wishful thinking clouded his vision. When Roberts called Eisner in February from Philadelphia's Ritz hotel, Eisner actually read his brush-off from a prepared script. Mitchell and his fellow directors publicly repudiated Roberts's bid, and stuck by Eisner even when a majority of Disney's voting investors, frustrated by the company's weak stock price, urged the Disney board to dump Chairman Eisner at its March annual meeting (held, coincidentally, in Philadelphia).

Did Comcast blow its chance? Could Roberts have gotten closer to its target if he had shown more patience instead of pushing the issue and all but forcing the Disney board to rally to Eisner and keep the unpopular chairman in power?[20] Comcast stock plunged 20 percent in the weeks after Eisner said no. For as long as the drop lasted, it did nothing to shine Brian Roberts's reputation as a dealmaker.

Critics like longtime telecom analyst Andy Kessler derided Comcast's Disney bid as a desperate move designed to save a faltering monopoly. From the day he announced the deal, Roberts strove to counter any such appearance. Again and again, he repeated, Comcast needed AT&T; it doesn't really need Disney. He and Burke could do a better job than Eisner, Roberts insisted, but they would happily go on about their business, fighting Murdoch and grabbing Internet customers away from AOL and other providers, if Disney demurred.

By April 28, Comcast finally tired of talking to itself and announced it was canceling the offer Disney had dismissed as soon as Brian had made it, two and a half months earlier. Roberts called his retreat an advance and blamed it on Disney's crazy behavior—on the Disney board's failure to act the way he expected when he called Eisner, man to man, and offered him $66 billion to go away.

"It seemed inconceivable to me that they would not want to have a dialogue with the world's premier distribution company," meaning Comcast, Roberts told investors and reporters summoned to a teleconference.[21] "To that extent," he admitted, "we miscalculated."

Like the Comcast executives of nearly twenty years before, outbid in an earlier expansion attempt, Roberts made a virtue of his failure: "We have always been disciplined in our approach to acquisitions. Being disciplined means knowing when it is time to walk away. That time is now."

To some observers, the conclusion was obvious: Young Roberts's reach exceeded his grasp. He really wasn't ready for the big time. "Brian Blew the Big One," taunted *Business Week*. Despite the family's boasted expertise with acquisitions, with Disney Roberts found himself "a little out of his league." He "misread his shareholders" and "badly bungled what should have been delicate negotiations with the Disney board," by relying on the Disney board, led by ex-Senator Mitchell, to turn on their chairman. Instead, "Mitchell was indignant, and any hope of dealing with the board evaporated." [22]

What would Brian do next? Maybe, he told reporters and investors, he would buy part of troubled Adelphia Communications, whose owners faced such a long list of criminal charges, partly for trying to build their own personal entertainment empire, allegedly under false pretenses, with shareholders money. Certainly, he would spend $1 billion buying back Comcast stock to prop up the share price.

The twisting nature of the story ensured that many in the media weren't taking Roberts at his word. What would Comcast really do? Buy MGM, maybe, or some other studio or network? Come back with a sweeter Disney offer, despite Brian's refusal to raise his price? Would he learn from this "miscalculation" and go on to the next bigger deal? In the face of changing technology and relentless new competition, would even a blockbuster deal really do Comcast any more good than big deals had done Michael Armstrong's AT&T?

Hello, Operator

Brian Roberts had timed the public part of his Disney adventure for the run-up to Disney's annual meeting, at the end of the mild winter of 2004. A successful bid, even a tough fight, would have dominated public notice of Comcast for months, maybe years, showcasing yesterday's cable monopolist as a rising and triumphant media power, providing high drama that could crowd aside any nagging questions about Comcast's slowing cable and Internet growth or customer-service troubles. But Disney's directors killed Roberts's offer so quickly, it was stiff before Comcast's own yearly meeting in May.

The dangers of such a situation are well known to professional spokespeople and other corporate apologists: The chief executive has failed in his latest grand design; the jumpy stock analysts and business reporters, feeling embarrassed they were swept up in the enthusiasm of a deal that didn't happen, look extra hard at his past record and extra skeptically at his next steps; it's a bad time for him to look confused, or even reflective.

But Comcast had survived other aborted mergers with brave talk and quick moves in other directions; Brian Roberts wasn't about to be caught looking like a possum in a speeder's high-beams. He shifted his message, and won new headlines, by telling investors all about the promise of Comcast's next big service: Internet-based telephone, a business Comcast had inherited from AT&T. Roberts promised that Comcast was preparing to offer cable phone service using voice-over Internet protocol (VoIP) to 40 million homes—including rival phone companies' most desirable customers. Never mind that Comcast, having inherited a million phone customers from AT&T, had actually been letting the business languish, with customers, at times, leaving faster than they were being signed up. It was, Roberts insisted, part of his current grand plan: "We have used this time since the merger to roll out the next generation of telephony that will include video, integrated messaging and all the things a computer can do," he told the shareholders assembled, once again, at Comcast's Wachovia Center in South Philadelphia.[23] Roberts delivered his pro-phone message, with Comcast's typical good timing, at a time when national media had suddenly woken up to cable's potential as a phone service and was ready to be led, yet again, to a sweeping corporate media vision.[24]

But whose vision? Comcast wasn't leading; in announcing it was now a phone company, it was following the other, smaller, bolder survivors in the cable business. "If you are a Cablevision subscriber, you're seeing it now," one Wall Street analyst told *USA Today*. "If you're a Time Warner customer, you'll hear about it by the end of the year. And if you're a Cox or Comcast customer, you'll hear about it in 2005."[25]

Still, this was one occasion where Roberts could stand with fellow cable companies against a larger foe. Instead of staying on the defensive against the phone companies' cut-rate Internet push, Roberts was putting Verizon and other big telephone providers on notice: They'd better worry about defending their dull old core phone service, or Comcast would take it away.

Too Soon the Prophet

Comcast is one of the rare companies in which the New York office is just a satellite to the Philadelphia headquarters, where the big decisions are made. New York is where C. Michael Armstrong had been left when he agreed to break up AT&T and join one of its buyers, Comcast. Nominally, he was chairman of Comcast, with corporate jet service and stock options and other regalia of power; in fact, he was a marginal player in the company's big decisions, ridiculed in the business press and Wall

Street telecom forums as the man who, having blown billions and de-stroyed AT&T, got kicked upstairs by the company that bought him against the day when he would bow out. "He should be remembered as the man who destroyed AT&T," Boston money manager Brian Bruce told Bloomberg LP as Armstrong prepared to retire.[26] No wonder he left Comcast a year early, stepping down at the 2004 annual meeting. Brian took the title of chairman from Armstrong.

But Armstrong's departure set off a revisionist history: In the end, hadn't Comcast prospered by adopting AT&T's multimedia vision? "Mr. Armstrong's strategy of combining cable and telephone was correct after all," Oppenheimer and Company analyst Thomas W. Eagan wrote.[27] For Eagan, at least, Armstrong's visionary attempt to combine cable and In-ternet "was scuttled by an impatient Wall Street": During the Internet frenzy, amid which Armstrong had built his expensive media empire, an-alysts had preferred the better-looking but ultimately "fraudulent" re-sults posted by media meteors like MCI Worldcom, instead of Arm-strong's more honest but less impressive-looking results at AT&T. "Unfortunately," Eagan noted, "timing was not on Armstrong's side." In-stead he was overtaken by Roberts, who had watched Armstrong plant his shareholders' money and the inheritance of AT&T's long but van-ished monopoly to grow a fabulous cable and Internet plantation. Brian had outwaited investors who were impatient for quicker results, swept in with a bargain offer for choice properties, and harvested where others had planted.

Vote Early and Often

Some of the professional corporate critics who tried to oust Walt Dis-ney's Michael Eisner in early 2004 went on to scold his would-be re-placement, Comcast's Brian Roberts, with even less success. A majority of the Disney shareholders who voted at Disney's Philadelphia conclave in March actually withheld their approval from Eisner; at Comcast's May meeting at the company's Wachovia Center in South Philadelphia, the votes weren't that close—except the last one.

Institutional Shareholder Services, a respected and independent firm that advises big nonprofits and pension funds about how to vote their shares, criticized Roberts's supervoting stock, which enabled him to wield one-third of Comcast's votes even though he owned less than 2 percent of the company; yet the firm still endorsed Roberts and his en-tire board.[28] A newer and more acerbic advisory firm, Glass, Lewis &

Company—which had met with Eisner and key Disney directors, pub- lished the entire conversation, and then issued a scathing anti-Eisner re- port—wrote a sixteen-page Comcast shareholder guide that urged in- vestors to split with Roberts on several measures on their yearly ballot. The firm opened its report by noting that Comcast "paid more [to execu- tives] than its peers, but performed worse than its peers," and got an "F" in Glass, Lewis's "pay for performance" rating. Glass, Lewis went on to urge investors to withhold their votes from longtime director Sheldon Bonovitz, who is managing partner of the Philadelphia law firm Duane, Morris, which works for Comcast, and who early in his career had mar- ried Suzanne Roberts's niece; and from S. Decker Anstrom, who, as president of Landmark Communications, had accepted $20 million from Comcast to carry the Weather Channel and related programming the year before. Yet, the firm also encouraged shareholders to approve Brian Roberts, his father, Julian Brodsky, and other Comcast loyalists who might as well have been family, finding no evidence of bias.

Despite its criticism of Roberts's pay, Glass, Lewis endorsed Com- cast's plan to pay another 4.3 million shares to senior employees; and agreed with Comcast's opposition to shareholder proposals that would impose an independent chairman, require contested board elections, and limit company political contributions. But the firm broke with Comcast to support a final proposal that would make all Comcast stock "one vote, one share." The goal of that resolution was to break the Robertses' special, controlling power over the company.

Like a feudal lord urging his peasants to back him as their representa- tive in a new elected government, Roberts appealed for votes. "The histor- ical success of Comcast is owed in large part to the respected and stable leadership provided by Mr. Roberts and before him by the leadership pro- vided by Ralph Roberts," the company told shareholders in a special letter sent five days before the vote, urging them to confirm his extra powers.

How much did the critics matter? Brian Roberts and his kinsman by marriage, Sheldon Bonovitz, were reelected with just under 95 percent of the company's voting strength. They outpolled Anstrom (of the Weather Channel) who got just 92 percent, while running slightly behind the rest of the directors, including Ralph Roberts, who all polled between 95 and 98 percent. Comcast also won the ballot questions, most with over 90 percent. The only ones that were close were the measure requiring a ma- jority of independent directors, which won about a quarter of the votes, and the one-share, one-vote proposal, which received 31 percent. Of course that accounting was under Comcast's current rules, with Brian wielding one-third of the vote despite his relatively small actual owner- ship. The votes for change shot up to 50 percent when Roberts's stake was factored out—and 55 percent if Roberts's ally Microsoft Corporation, Comcast's largest single owner, was discounted, according to the tally by the Communications Workers of America, struggling to organize Comcast employees.[29]

Outside in the parking lot, dozens of Comcast electricians and installers rallied for "Comcast Subscribers and Workers Rights."

"What time is it?" Pennsylvania AFL–CIO President Bill George yelled at the crowd.

"Union time!" the visitors shouted back.

"We're nothing to them," George said of Brian Roberts, David L. Cohen, and other high-ranking Comcast officials. But, he exhorted members as they approached the hall for the annual meeting, "be respectful."[30]

Spammer, Brawler, Thieves

If Comcast's low profile ever protected it from scrutiny, the charm wore off sometime between the AT&T takeover and the failure of the initial Disney bid. In May 2004, Washington, D.C., reporter Declan McCullagh quoted a Comcast engineer, by name, telling an "antispam working group" that "we're the biggest spammer on the Internet," adding that of nearly 1 billion Comcast e-mail messages sent each day, the majority "are coming from compromised machines," whose owners can't figure out how to prevent them from being abused.[31] The quote was gleefully repeated in Web logs and trade journals, adding little luster to Comcast's reputation.

Across the country in Los Angeles, the sparkplug head of TV gossip network (and Comcast–Disney joint venture) E! Networks, suburban Philadelphia native Mindy Herman, was reported headed for "an abrupt but lucrative end" in that job. A *Los Angeles Times* story had Herman raiding employees' valuable gifts from story subjects (which apparently entertainment reporters feel entitled to keep), appropriating company property, and engaging in "hand to hand combat" with an unidentified woman at a Hollywood burlesque club. Comcast, E!'s managing partner, declined comment. The *Times* put Herman's expected severance, based on the channel's expansion on her watch (much of it due to Comcast acquisitions), at a cool $20 million.[32]

And in Florida, Comcast reported itself shocked to learn that as many as one-third of cable households were pirating the signal for free. The company responded with a sober round of infomercials, threatening to catch freeloaders and reminding them that cable theft "can collectively affect communications, including those used by aircraft."[33]

In its ads, Comcast apparently made no mention of its own record fine for doing exactly that—interfering with air traffic signals—over the Midwest twenty years before, an accusation Comcast officials had publicly disputed and privately ridiculed. But even the suggestion, by Comcast, that a hotwired cable box could send jets tumbling into Biscayne Bay was too much for the ironists at the *Broadband Reports* online newsletter, who published a brief item on Comcast's attempt to link numerous social ills to ripping off your cable company, under the headline, "It could crash a plane, you know."[34]

The Tony Soprano of Cable?

As Comcast planned its annual meeting, RCN, the Boston-based cable company that had tried to compete with Comcast in Philadelphia and dozens of other markets, was preparing to file for Chapter 11 bankruptcy protection from its creditors. Like Mike Armstrong's AT&T Broadband, RCN had gone through billions of dollars in other peoples' money, only to fall spectacularly short of what investors had been promised. Lashing out at cable in the *Wall Street Journal*, RCN chief executive David C. McCourt laid the blame for his apparent failure at Comcast's doorstep—then predicted ultimate victory for RCN and other challengers to the established order.[35]

"I believed—and still do" that competing home communications systems can "deliver every type of service that consumers would want. Not just phone, cable and Internet, but home security, energy monitoring," even appliance repairs, McCourt began.

He praised the competition that developed between the once-monopolistic phone companies after deregulation in the 1990s. But "by contrast, the major companies still divide-and-rule territory," McCourt wrote. Their "local fiefdoms" use "the same businesss model championed by Tony Soprano," the cable TV mobster. "They force you to rent their proprietary converter boxes," instead of letting customers buy standard boxes at the store. They force consumers to buy channels they don't want to get the ones they do. "Can you imagine if Wal-Mart insisted that you buy a hairbrush when all you want is a toothbrush?" Comcast plus Disney would have made Disney programs "harder to get, more expensive, or less convenient."

McCourt's hope: that bankrupt companies like RCN, shedding their debt, will come roaring back to use the new voice-over Internet protocol technology against the cable companies, by providing cable services through phone lines. "That possibility should worry today's cable giants who have been ignoring the logic of economics, the possibilities of technology and the interests of consumers for far too long."

How Cable Should Have Been

Comcast's marriage of cable with the World Wide Web is subversive of an old Internet conceit: that the Net will somehow make big media obsolete. Instead, companies like Comcast, by controlling the distribution of information, enjoy an increasingly lucrative, high-profit business by charging a toll to use the Net.

Should it be different? Could it be? Howard J. Blumenthal, a TV veteran and visionary/pitchman of a type more familiar in Los Angeles and New York than anywhere else, sat at a tiny table in an old-fashioned fish restaurant in Center City Philadelphia, the kind of place where natives gather for regional specialties like shad roe fried in fat and snapping-turtle soup laced with sherry. It was late winter, the beginning of shad season, and the place was packed.[36]

Blumenthal was, like Roberts and Burke, born into TV. His father ran the old game show, Concentration. Blumenthal wrote the business plan for MTV and helped start Nickelodeon; won national awards for his public-television kids' geography show, *Where in the World Is Carmen Sandiego?*; has written a library shelf of syndicated columns and books (*This Business of Television*); and was for a time a senior marketing officer at Philadelphia's one really successful Internet company, the music trader CDnow, whose purchase and shutdown by the German conglomerate Bertelsmann meant a loss for metro Philadelphia's Web-based economy, but a fine payday for Blumenthal's employers.

Finished with his gravlax, Blumenthal popped his laptop open; his new Web site, newcenturytv.com, was streaming away, clear and fast, for his lone companion. Click, here was his travelogue of Copenhagen. Click, there was BAT Channel, on which an engineer soft-sells his company's custom sound systems. Click, an attractive young woman was walking prospective students through LaSalle University. Click, there was a prototype channel for Rutgers University. These were among his clients. For a modest fee, Blumenthal said, anyone can now enter the video production and distribution business.

"Everyone that wants a channel can have one now," Blumenthal said happily. "Three thousand channels." Or three hundred thousand—with varying degrees of professional polish—plus millions of amateurs. "This is what cable TV was supposed to be. This is the Internet today, and to-morrow."

But does that mean the Internet will kill television as we know it? Will it make all those cable TV systems obsolete? Blumenthal shook his head. Sure, he said, there's room for creative guys like him to set up Web media for anyone willing to pay. But "you still need big companies for the big stuff." Disney's copyrights represent an awesome power; the company isn't about to let everyone download its stuff from the Web for free. Little channels like his might fill profitable or useful niches, but media companies, Blumenthal said, will still control the most lucrative uses of the small screen.

And you can't view these wonderful new downloads on a kludgy old dial-up Internet service; or, for the most part, by wireless connections—at least, not yet. The channels linked to newcenturytv.com and its clients' Web sites, for all their liberating appearance, depend on broadband Internet and its big corporate providers. Blumenthal is there because the cable companies built the way; he depends on them, and his viewers pay for it in their monthly cable bills. What's good for those liberating little channels is great for Comcast and the other cable providers. So don't look for revolution here: It is in Blumenthal's interest, he said, for even fast-moving independent operators like him to "have a good relationship with Comcast."

Shame on Us

Back at Comcast Investors' Day in May 2003, Brian Roberts had tried to sum up where he felt he'd brought his father's company with its expensive mergers and rapid growth. "We're in a very unique position," he boasted. "We can bring to the table, not just money, but twenty million homes." Cash flow was up; profit margins ought to keep rising.[37]

Asked again about Comcast's high prices, Roberts said he believed the company had bought customer loyalty with its high-speed Internet service. "I don't think people will give up e-mail and everything they've stored," just for cheaper satellite TV, he said.

Won't Comcast ever back off from its imperial ambitions, at least for investors' sake? Would it ever ease up on the deal making, work its businesses, and show a profit?

"It's a high class problem," Roberts demurred. There are many claims on Comcast profits: "We have debt to pay off. We have stock to buy back."

To what end? "This company will be in a position to look at operations that are not available to the regular investor," Roberts affirmed.

"Shame on us if we take our eye off the ball."

NOTES

INTRODUCTION

1. Recordings and transcripts of Comcast investor presentations are available from a variety of sources, including Comcast's investor relations site (www.cmcsa.com) and the Fair Disclosure (FD) Wire, published by CCBN, Inc., and its subsidiary, FDCH e-Media, Inc. (formerly known as Federal Document Clearing House, Inc.). Comcast's quarterly conference calls are open to investors, the news media, and other interested callers.

2. Brian L. Roberts, interview by the author.

3. A typically Philadelphia ambivalence is expressed by business writer Dan Rottenberg, biographer of A. J. Drexel and former editor of a free Philadelphia weekly before it was taken over by Brian Roberts's brother-in-law: "You have to admire [the Robertses] for still coming at this like they're first-generation immigrants." Yet he finds their approach to business little short of predatory: "Again and again you see they tilt the playing field to avoid competition."

4. Daniel Aaron, with David A. Long, *Take the Measure of the Man: An American Success Story* (Philadelphia: Veritas Publishing, 2001), 135; Ralph Roberts wrote the foreword. Aaron, who died in 2003, was Roberts's first hire in the cable TV business that became Comcast.

5. Ted Hearn, "Ops Blame Nets, Seek Tier Help," *Multichannel News*, periodical published by Reed Elsevier, 12 May 2003.

6. Pat Croce, interview by the author.

7. Brett Pulley, "Comcast's Muscle Man," *Forbes*, 4 April 2004.

8. Julian A. Brodsky, interview by the author.

9. For a vivid account of Roberts's sneak attack on Malone's company and Malone's infuriated reaction, see Mark Robichaux, *Cable Cowboy: John Malone and the Rise of the Modern Cable Business* (Hoboken, N. J.: John C. Wiley & Sons, 2002), 218–219.

10. Brodsky.

11. Reed E. Hundt, *You Say You Want a Revolution: A Story of Information Age Politics* (New Haven: Yale University Press, 2000), 55.

12. John Helyar, "The First Family of Cable," *Fortune*, 29 October 2001.

13. Center for Digital Democracy, "FCC's Approval of AT&T Comcast Merger Leaves Cable TV and 'Net Users Vulnerable to a Media Mega-Monopoly," press release, 13 November 2002; see also, CDD, "What Brian Roberts of Comcast Doesn't Want You to Know about the Future of Cable and Internet Competition," 1 November 2002, and "No Competition: How Monopoly Control of the Broadband Internet Threatens Free Speech: An ACLU White Paper," 13 November 2002.

14. J. D. Power and Associates, "2003 Residential Cable/Satellite TV Customer Satisfaction Study," 19 August 2003. University of Michigan Business School's National Quality Research Center, and partners, in "ACSI: Customer Satisfaction Rebounds Sharply," 21 May 2003, found that, among sixty major service companies, cable TV enjoyed the worst reputation with consumers. See also, for example, Thomas W. Watts, "Will Satellite TV's Growth Run Out Soon?" securities research report, S. G. Cowen & Company, 18 June 2003, a survey of 2,700 TV viewers that reported satellite TV users "much more satisfied" with both value and customer service than cable users.

15. See, for example, Morton Bahr's address to Communications Workers of America, District 3 conference, 25 March 2002, Miami, Florida. The author is covered by a CWA collective bargaining agreement.

CHAPTER 1

1. Like other Roberts stories, this has evolved in the retelling. The version cited was published in Michael L. Rozansky, "The Cable Guys: Comcast's Brian and Ralph Roberts Make Their Big Play; Wired to the Future, Tomorrow's TV: 500 Channels, a Million Choices and Two Quiet Cable Tycoons," *The Philadelphia Inquirer,* 11 August 1996. Roberts later called his garden depredations "the first *business* thing I remember" (emphasis added), in Kathi Ann Brown, *Wired to Win: Entrepreneurs of the American Cable Industry* (Fairfax, Va.: I. Martin Pompadur, Spectrum Publishing Group, 2003). See chapter on Ralph Roberts. The story Ralph tells in Brown's book expands in some detail the comic and poignant personal incidents he had recounted in previous newspaper and magazine articles.

2. The Roberts family declined to talk about its origins, and senior Comcast officials claim ignorance. For example, Julian Brodsky told the author that in forty-one years of intimate collaboration with Ralph, he doesn't remember the subject coming up. If true, the omission could imply a certain distance between Roberts and even his most trusted lieutenant; it also prefigures Brian Roberts's preference for personal privacy, so different from the public display practiced by his mother and previous Philadelphia Fleishers.

Thanks to Robert Foley, a cousin of the author, of Cortlandt Manor, New York, and his firm, WriteFocus Communications, for his research into Roberts sites, articles, and documents in New York. The New York State death certificate of Robert M. Roberts, Ralph's father, states he entered the United States together with his own father, listed as Max Roberts, and his mother, the former Rebecca Stomansky, twenty-nine years before the younger man's death in February 1933. Max's tombstone in a Jewish cemetery outside New York names him in Hebrew characters as "Mordecai, son of Ruvayn," in English as Max Roberts, in the custom of the time.

According to the same document, Max, Rebecca, and Bob Roberts were all born in "Diobitzer, Russia." Gazeteers list no such place. But according to the system developed by genealogists Randy Daitch of Los Angeles and Gary Mokotoff of New Jersey to compare place and personal names in Eastern Europe, "Diobitzer" exactly matches the formerly Russian-controlled, Jewish-majority town of Dubossar (later Dubasari) in the ex-Soviet republic of Moldova. Robert Roberts's later service as treasurer of the Hebrew National Orphan

Home, founded by Moldovan Jewish immigrants a few years after his arrival in New York, also suggests roots in that community. The author thanks Deena Schwimmer of the American Jewish Historical Society for her guidance.

Old accounts show Dubossar as a local center where even peasants were Jewish, a place with vigorous squabbling factions—Hasidic, Zionist, socialist, and capitalist—noted popular (klezmer) and classical musicians, and a Jewish militia that strove to discourage the kind of murderous anti-Jewish riots that bloodied Moldova in 1903 and again in 1905. The Robertses left the country soon after the first and worst attack, which killed fifty in the nearby Moldovan capital of Kishniev (Chisnau); investigators called it a mob response to the sensationalized but unrelated deaths of a Christian woman and child in Dubossar. For the next twenty years Dubossar Jews continued moving to New York, until both Soviets and Americans shut the borders. German invaders massacred many of Dubossar's remaining Jews in 1942.

3. Westchester Social Record of 1930, Roberts family listing.

4. Brown, *Wired to Win*, 264.

5. Fitzgerald set the start and end of his 1919 short story "Myra Meets His Family" at the Biltmore; he portrays the vast lobby as a hangout for preppy college-boys and fortune-hunting girls; like the other hotels in the Grand Central neighborhood, it was also a meeting place for politicians, speculators, and other operators who fed off the city's Jazz Age prosperity in the last fat years of the Tammany Hall political machine.

6. "300 at Funeral for R. M. Roberts at His Late Home," *New Rochelle Standard-Star*, 20 February 1933.

7. Brown, *Wired to Win*, 265.

8. Rozansky, "The Cable Guys."

9. Aaron, *Take the Measure of the Man*, 56. Daniel Aaron was a popular man whose contemporaries credit his energy and innovation for much of Comcast's early success. The book mixes Comcast history with Aaron's dramatic personal story: His parents, refugees from Nazi Germany, killed themselves, and he was raised and later nurtured by a series of Jewish patrons, culminating with Roberts. Julian Brodsky has disputed a few points in Aaron's account.

10. In stressing the Comcast founders' common roots, Aaron, the onetime Philadelphia *Bulletin* reporter, showed the journalist's tendency to search for patterns. But Brodsky, the skeptical accountant, called any attempt to make too much of their parallel past "quite a stretch"; for him it's coincidence, not destiny.

11. Cable News Network, transcript 99051600V39, 16 May 1999.

12. See, for example, Karen Heller's amused but ultimately quite respectful profile of Suzanne, "Raring to Go," *The Philadelphia Inquirer*, 19 October 2003, which also pictures the blond octogenarian in a biker's outfit at a Harley-Davidson dealership and theme-restaurant near Comcast's studio in out-of-the-way New Castle, Delaware.

13. Highlights of the first hundred years of the Fleishers in Philadelphia are well documented in the city's newspaper archives, including *Philadelphia Inquirer* records from the 1920s through 1950s, which the author reviewed; as well as in such generally flattering community histories as Henry Samuel Morais, *The Jews of Philadelphia* (Philadelphia: Levytype Co., 1894/5654 [Jewish year of publication, as listed in original edition]).

14. Like the works of Alfred and Samuel Fleisher, Selma Gerstley Fleisher Sunstein's long family, social, and charitable career is documented in the news research archives of the *Philadelphia Inquirer*.

15. Carole Saline, "The Cable Gal," *Philadelphia Magazine*, March 2001.

16. Ralph Roberts, remarks on accepting the American Horizon Award from the Media Institute, 23 October 2002.

17. Rozansky, "The Cable Guys."

18. Ibid.

CHAPTER 2

1. Ralph Roberts described his work at Aitkin Kynett and for Senator Benton to the *Inquirer*'s Peter Binzen, "Comcast Boss Has Taken the Roundabout Way Up," *The Philadelphia Inquirer*, 21 July 1985; and again to Kathi Ann Brown, *Wired to Win: Entrepreneurs of the American Cable Industry*.

2. Prescott Bush won Connecticut's other Senate seat in a later election. Of course, the Bush dynasty's careful blend of oil, corporate, and political influence established a power far greater than Benton's.

3. Benton isn't the only media thinker with personal ties to the Roberts family. University of Chicago law professor Cass Sunstein, a cousin-by-marriage of Brian L. Roberts through his grandmother Selma Gerstley Fleisher Sunstein, is the nation's best-known critic of the polarizing and divisive tendencies of modern electronic media. His thesis is that people are using their new digital media choices, not to sift new and conflicting ideas, but to confirm their own ever more extreme interests and prejudices. Contacted by the author, Sunstein confirmed his Philadelphia ties but demurred when asked whether he and Brian have ever talked about what Comcast can or should do, as an Internet provider and media giant, to improve the nation's fragmenting political culture.

4. Although Ralph Roberts has given other versions that vary in details since his retirement, I am generally following Binzen's earlier account here.

5. Joseph Epstein, "Just Plain Bill," *The New York Review of Books*, 22 October 1970.

6. Ralph Roberts recounted his Pioneer career and his experience with Philadelphia banks to the author in a 1997 interview. Daniel Aaron's book, *Take the Measure of the Man*, also contains a detailed treatment, and the outline forms a standard part of numerous Roberts profiles in the Philadelphia and national business press as well as contemporary items in the *Philadelphia Inquirer*.

7. Dan Rottenberg, *Wolf, Block, Schorr & Solis-Cohen: An Informal History 1903-1988* (Philadelphia: Wolf, Block, Schorr & Solis-Cohen, 1988), 55.

8. Aaron, *Take the Measure of the Man*, 131.

CHAPTER 3

1. "Video's Effect on Political Campaigning," an article in Walter Annenberg's *TV Digest*, 14 June 1952, gave "the strikingly attractive Suzanne Roberts" and her book a fawning review, citing the victorious Philadelphia Democrats' "dependence" on her campaign TV shows. The review opens with a quote from her book: "No cold fish need worry about being a dead fish on television." The

original caption notes, in the interest of full disclosure, "Mrs. Roberts' book is published by *TV Digest*." (See first picture in photographic insert in this book.)

2. Mahanoy City, Pottsville, and other Schuylkill County towns were a cradle of cable TV, not just because they couldn't get broadcast, but also because they could afford the alternative in the afterglow of wartime mine and factory orders. It's been downhill ever since. The region's upper crust was portrayed with little flattery and less mercy in the best-selling novels of prolific native son John O'Hara, but he moved to New York before cable barons could join his catalog of ambitious, self-indulgent, and sexually frustrated provincial aristocrats. The county has aged badly, and in the rundown boroughs savaged by O'Hara, now little read elsewhere, he has become a cottage industry, now that he's safely dead.

3. Aaron put these words in Julian Brodsky's mouth, *Take the Measure of the Man*, 135. Brodsky can be as blunt and funny as any member of that candid generation of cable pioneers he belongs to. But when asked about Aaron's quote comparing early cable economics to theft, in interviews with the author at Comcast corporate offices, Brodsky insisted he "would never, ever have said anything of the kind. . . . I might have said 'the greatest thing since sliced bread.'" Brodsky also winced when reminded of Roberts's own self-deprecating stories, such as his golf-club misadventure, detailed in Chapter 1: "Why does he have to tell that one? Do you think that makes him look good?"

4. David Lieberman, "Father-son Odd Couple Make Bid to Rule Cable, One's Calm, One's Wary, Together They're Comcast." *USA Today*, 23 July 2001.

5. Julian Brodsky, interview by the author.

6. The Cable Center in Denver, Colorado, where some of the nation's largest cable companies were formerly located, offers an online archive of cable history and individual accounts by cable pioneers, including John Walson, at www.cablecenter.org.

7. The Necho Allen was a fancy backdrop in John O'Hara's Pennsylvania stories, just as Manhattan's Biltmore was for Fitzgerald. The Biltmore is now a bank tower; the Necho Allen is senior-citizens' housing. "O'Hara was the real Fitzgerald," New York writer Dorothy Parker once said.

8. From a 1986 interview recorded by the Cable Center. A short history of the company is also provided on Web sites of Security Electric's successors, including www.secv.com and www.sectv.com.

9. Mary Alice Mayer, *Oral Histories: John Walson, President/Chairman, Service Electric Corp.*, The Cable Center, August 1987.

10. Milton Shapp's career is documented in the *Philadelphia Inquirer*'s archives, though 1960s coverage was colored by then-publisher Walter Annenberg's intense dislike of his sometime equipment supplier. In *Take the Measure of the Man*, Daniel Aaron, who worked for Shapp, paints a vivid picture of Shapp's role in expanding the industry in the 1950s, which is supported by Cable Center accounts. Mark Robichaux's biography of John Malone, *Cable Cowboy*, 29–31, cites Malone's account of Shapp's former company in disarray after the founder left to run for governor of Pennsylvania. Tyco International chairman Ed Breen, former chief executive of Jerrold Electronics' successor General Instruments, and Frank Drendel, chief executive of cable maker Commscope Incorporated, were among the old Jerrold hands who helped fill in the company's later history, which comes back into this story as Ralph Roberts starts calling for digital services in the early 1980s.

11. K. C. Neel, "Hall of Fame Honorees," *Cable World*, PBI Media/Gale Group, 3 November 2003.

12. Robichaux, *Cable Cowboy*, 29–31.

13. Julian Brodsky, interview by the author. "Warren "Pete" Musser's speculative investment company, later known as Safeguard Scientifics, played a recurring role in the Comcast story. In 1963 it sold Ralph Roberts his first cable system; in the 1980s Musser convinced Comcast to invest with Safeguard in QVC, getting Comcast into the television programming business; in the late 1990s he allowed Comcast and senior Comcast officials, including Julian Brodsky, to profit briefly but handsomely from the Internet bubble by purchasing shares early and cheap in a Safeguard-backed company called Internet Capital Group. Meant to be "the General Electric of the Internet," Internet Capital was briefly worth more than General Motors—$50 billion plus—despite its conspicuous lack of profits. The share price plunged with other Internet stocks in 2000.

14. Chestnut Street, where Daniel Aaron, in his memoir, *Take the Measure of the Man*, says the meeting took place, was Philadelphia's premier shopping district at the time. But Julian Brodsky said the encounter may actually have taken place a block south, on Walnut Street, the city's old financial axis. Neither site has been fitted for a brass plaque. Presumably, though, Comcast would prefer that its civic contributions were recognized with something like the tax break the Hon. Ed Rendell, Pennsylvania's governor and Comcast talk-show host, has campaigned to secure the company, for a new headquarters site it is mulling a few blocks away.

15. The sale was noted in the *Philadelphia Inquirer* at the time. Writing in the paper years after the fact, Peter Binzen, "Comcast Boss Has Taken the Roundabout Way Up," *The Philadelphia Inquirer*, 21 July 1985, and Michael J. Rozansky, "Hooking Up a TV Empire," *The Philadelphia Inquirer*, 20 November 1994, give more detailed accounts; Aaron adds color. Binzen is particularly scrupulous in recounting financial arrangements and Ralph Roberts's insistence on voting control. The author also reviewed these events with Julian Brodsky, who confirmed Binzen's retelling, which gives more emphasis to the roles of Joseph Roberts and Fred Wolf than is found in more recent accounts.

16. Binzen, "Comcast Boss."

CHAPTER 4

1. Julian Brodsky, interview by the author. Daniel Aaron has a similar version.

2. Daniel Aaron, *Take the Measure of the Man*, 139–146.

3. In his memoirs, 141, Aaron claims the building was called "three foot" because its location, three feet across a municipal boundary, gave it a special tax status. Brodsky said Aaron is mistaken, and since Brodsky was the accountant, the author has used his version here.

4. For more on Theodore Aronson, see Chapter 8, 65–71. Aaron describes the relationship among Roberts, Brodsky, and himself in his Chapter 13, "The Boys," *Take the Measure of the Man*, 151–165. Brodsky and Roberts have given similar accounts; but however cooperative the triumvirate was at times, there

were other key players—Fleisher and Roberts family members and friends in the 1960s, professional managers after that—and Roberts was ultimately the boss.

5. The push to make insurers more aggressive investors had a huge impact on U.S. business over the next generation, for better and worse. It was led by a small but influential 1960s investment-banking firm, Cogan, Berlind, Weill & Levitt, whose partners included Arthur Levitt, future chairman of the Securities and Exchange Commission, and Sandy Weill, creator of Citigroup. In the 1990s, Weill's persistence and Levitt's acquiescence made possible the final removal of the barriers to corporate financial speculation dating to the Great Depression, a huge change with hugely mixed results. For its part, Comcast scrupulously avoided default, and the insurers' investment was an unmixed blessing.

6. Julian Brodsky, interview by the author.

7. Aaron, *Take the Measure of the Man*, 153.

8. Harold Fitzgerald "Gerry" Lenfest, interview by the author.

9. Julian Brodsky said he doubts Roberts would have moved his then-cable chief, Daniel Aaron, to accommodate Lenfest.

10. See Chapter 6, 53–58.

CHAPTER 5

1. Dan Rottenberg's *The Man Who Made Wall Street* (Philadelphia: University of Pennsylvania Press, 2001) presents Philadelphia's Anthony J. Drexel as the mentor for J. Pierpont Morgan, rather than the conventional wisdom that young J. P. learned at the knee of his father Junius in England. The house of Drexel, Morgan & Company, an ancestor of the giant J. P. Morgan & Company and more directly of the infamous Drexel, Burnham, Lambert junk-bond house of the 1980s, was at first based in Philadelphia (as were the two predecessors of that other global giant, Citigroup Smith Barney). According to Rottenberg, early in their partnership Morgan made weekly trips to Drexel's office in Philadelphia to report to his senior partner. Even after stock trading had migrated from Philadelphia's bourse to the New York Stock Exchange, Philadelphia financiers kept a certain eminence as leading sellers of streetcar, Mexican War, and Civil War bonds. No major investment or commercial banks call the Philadelphia area home today, though it still houses a beehive of private and midsized public investment firms, as well as the suburban Vanguard Group, the world's largest mutual fund company. But individual Philadelphians—Michael Milken, first hired by Drexel in Philadelphia; Jack Grubman, telecom analyst; Frank Quattrone, Internet dealmaker—have been prominent in the Wall Street scandals and controversies of the 1980s and 1990s.

2. Leon Sunstein Jr., interview by the author.

3. Julian Brodsky, interview by the author. These events are also described in Daniel Aaron's *Take the Measure of the Man*, 161, and in contemporary news accounts.

4. Noel Stansell, e-mail correspondence with the author, February 2004.

5. See, for example, Grace Madley, "One Family's Vacation . . . as Volunteers," *The Philadelphia Inquirer*, 17 November 1971.

6. "Hall of Fame 2000 Members," The Cable Center, Denver, Colorado; announcement on Roberts's joining the Cable Hall of Fame.

Chapter 6

1. Allen Sloan, "Bring Plenty of Money," *Forbes*, 10 December 1979. A knowing and prescient early account of the national cable TV business.

2. The Cable Center interviewed Sidney Topol in 1980 and posted the transcript on its Web site, www.cablecenter.org.

3. Mark Robichaux, *Cable Cowboy*, shows the impact of the Frazier–Ali fight in convincing local cable operators to sign up for HBO and other national networks.

4. Julian Brodsky, interview by the author.

5. Sloan, "Bring Plenty of Money."

6. Brodsky, interview by the author.

7. Ibid.

Chapter 7

1. Harold Fitzgerald "Gerry" Lenfest, interview by the author.

2. Comcast's 1977 annual report shows cable used 74 percent of Comcast's assets but yielded only 48 percent of Comcast's sales, with Storecast and Muzak giving the rest. But cable profits as a percentage of the total had risen, from 43 percent in 1973, to two-thirds by 1977. The trend toward cable was clear, but still far from complete.

3. Dan Hulbert, "Cable TV Companies Standing By," *The New York Times*, 23 November 1980.

4. Choosing Ballard, Comcast once again caught a rising star. Prominent Ballard lawyers included Philadelphia mayor turned Pennsylvania Gov. Edward G. Rendell, who has honored the company's Philadelphia loyalty and campaign contributions by hosting a Comcast sports show and fighting for Comcast tax breaks, and Rendell's ex-chief of staff, David L. Cohen, the brainy architect of Comcast's political and public relations strategies since he joined the company in 2001.

5. "The Surprising Success Stories in Cable Television," *Business Week*, 12 November 1984.

Chapter 8

1. Tom Lowery and Amy Barrett, "A New Cable Giant," *Business Week*, 18 November 2002.

2. Karen Heller, "The Bush Business Is Politics," *The Philadelphia Inquirer*, 31 July 2000.

3. Michael Bamberger, "Squash Lessons," *Sports Illustrated*, 4 December 2000.

4. Julian Brodsky, interview by the author.

5. Steve Goldstein and Akweli Parker, "The King of Cable," *The Philadelphia Inquirer Magazine*, 23 June 2002.

6. Christopher Stern, "A Rare Miss for Roberts, Comcast's Dealmaking Executive Surprised by Disney's Rebuff," *The Washington Post*, 29 April 2004.

7. *CableFax*, PBI Media/Gale Group, 13 May 2003.

8. Thanks to Paul F. Miller, former Drexel chairman, for his background on the firm and its evolution.

9. When Brian became head of Comcast in 1990, "I wondered. He was young," Leon Sunstein Jr. told the author. "But he knew the trade. . . . I remember Ralph telling me at one point, 'You know, we were going to go out and look for somebody, but I couldn't have gotten somebody better.' He trained him; he was awfully impressed with what he did."

10. Brodsky, interview by the author.

11. Theodore Aronson, interview by the author. Aronson is among the most erudite of professional money managers; though he's not a man to offer casual stock tips, his generosity in explaining how the investment business works has endeared him to young stock analysts and reporters.

12. Norman Black, "FCC Fines Michigan Cable TV," Associated Press, 24 April 1981.

13. See Chapter 11, "What's Next," 89.

14. Ralph Roberts, "Brian L. Roberts Named President of Comcast Corp.," Comcast press release, 7 February 1990.

15. Beverly Schuch, "The Unlikely Story of Ralph Roberts' Billion Dollar Media Empire," Cable News Network, 16 May 1999.

Chapter 9

1. Julian Brodsky, interview by the author.
2. Edward Breen, interview by the author.
3. Comcast press release, 22 October 1984.
4. Frank Drendel, interview by the author.

Chapter 10

1. Comcast press release, 19 April 1983.

2. "Fight TV at Shea Barred," *The New York Times*, 13 April 1983.

3. Bogner recorded a respectable record of 15-2-1 before retiring from the ring in 1984. His later comeback was not successful.

4. The *Philadelphia Inquirer* covered the city's cable franchise woes extensively from the 1960s until the business was finally awarded in the 1980s.

5. Comcast press release, 22 October 1984.

6. See, for example, Comcast's 1982 annual report. The company reported little or no profit as it expanded its purchase of other cable systems in the middle 1980s.

7. Geraldine Fabrikant, "Comcast's Consistent Profits," *The New York Times*, 23 July 1985. Brodsky was speaking after the Storer deal, as well as the Baltimore-area acquisition.

8. Drew Lewis made a career of reorganizing companies, even whole industries, and profiting from the changes he'd wrought. Active in the long, lucrative liquidation of Philadelphia's bankrupt Reading Company in the 1970s, Lewis in the late 1990s made another fortune running and finally selling Union Pacific Railroad after lobbying successfully for relief from old merger restrictions. The deal was a disaster for railroad shippers, but capped Lewis's successful career in both politics and business.

9. The arrangement would later be modified to keep so-called "basic" cable rates within certain broad limits.

10. David A. Vise, "Storer to Weigh Bid by Cable Firm," *The Washington Post*, 18 July 1985.

11. Fabrikant, "Comcast's Consistent Profits."

12. "Comcast's Boss: Building One Step at a Time," *USA Broadcasting*, 3 February 1986.

CHAPTER 11

1. Philadelphia's pursuit of cable from 1963 to 1986 was the subject of many articles in the *Inquirer* archives, which are adapted freely in this chapter.

2. William Robbins, "Philadelphia's Mayor and Council Clash over Cable Television," *The New York Times*, 26 September 1983.

3. Mark Robichaux, in *Cable Cowboy*, provides some fine examples of the demagoguery employed by Senator Gore against the combative John Malone and other cable pioneers over their questionable business and customer practices. Ralph Roberts worked closely with Malone but was careful not to bait the politicians, or face them alone.

4. Julian Brodsky, interview by the author.

5. "Cable Television: New Punters in the Cultural Wasteland," *The Economist*, 25 October 1980.

6. "Comcast's Boss: Building One Step at a Time," *Broadcasting*, 3 February 1986.

7. Don Steinberg, "The Man Who Dared to Take on the Mouse," *The Philadelphia Inquirer*, 15 February 2004.

CHAPTER 12

1. "The Candy Man (Can)," from the musical show based on the late Roald Dahl's *Willy Wonka and the Chocolate Factory*. Oddly, the parody may come closer to Dahl's sardonic style, which portrays the majority of adults as cheerfully and demonically selfish, than the unreflectively happy show song on which the parody is based, a prospect that must comfort Dahl's grim-comic soul.

2. Steve Donohue, "Ralph Roberts: Comcast Keeps the Faith," *Multichannel News*, 9 June 2003.

3. Comcast Corporation conference call to detail merger discussions with Walt Disney, official transcript, 12 February 2004.

4. Mark Tran, "Diller's Media Ambitions Foiled Again by Surprise Spoiler from Dynamic Duo," *The Guardian*, 18 July 1994.

5. Rose De Wolf, "Success Channel," *Philadelphia Daily News*, 15 July 1994.

6. The Cable Center, announcement of Ralph Roberts's induction to the center's Hall of Fame in 2000. Comcast would buy a majority stake in QVC.

7. Michael Rozansky, "Comcast Went from Backing Diller to Buying His QVC," *The Philadelphia Inquirer*, 14 July 1994.

8. Shareholders dislike corporate mergers because most of them don't work; they waste assets and stress workers, vendors, and customers. Bondholders hate them even more because they tend to increase debt. CEOs, investment bankers,

Wall Street lawyers, and stock traders, on the other hand, love mergers, the first three groups because they are richly paid per deal, and the traders because they tend to make money whenever stocks change hands. The large and increasing number of mergers in recent years shows how much power professional Wall Street and corporate managers enjoy, and how little belongs to stock and bond investors.

9. Steve Goldstein and Akweli Parker, "The King of Cable," *The Philadelphia Inquirer Magazine*, 23 June 2002.

10. Ibid.

11. Kate Oberlander, "Comcast Deal Would Connect Cable, Cellular," *Crain's Electronic Media*, 13 May 1991.

CHAPTER 13

1. Jon Lafayette, "Now That's a Home Field Advantage," *Cable World*, 2 September 2002; Jeff Gelles, "Sports Fans Can Offer a Preview of Merger," *The Philadelphia Inquirer*, 18 February 2004. Gelles, in his column, warned about "the dangers of any company's having too much power in its niche." Comcast says it is following the law and has so far prevailed in lawsuits by DirecTV and other satellite companies.

2. Charles Paikert, "Working Class Hero," *Multichannel News*, 11 March 2002.

3. Lafayette, "Now That's a Home Field Advantage."

4. Nevertheless, Comcast is trying to centralize local news, with its CN8 studio on the flats south of Wilmington, Delaware, replacing far-flung facilities up and down the East Coast.

5. The development of Comcast–Spectacor was covered in a number of articles in *The Philadelphia Inquirer*, 1995–2003.

6. Pat Croce, interview by the author.

7. William C. Smith, "Building a Cable Behemoth," Arthur Block profile, *National Law Journal*, 26 May 2003.

8. Ben Wanger, "Academy Edition," *Germantown Academy Newsletter*, 22 February 2001.

9. Phil Sheridan, "Jack Williams Altered the TV game," *The Philadelphia Inquirer*, 13 April 2001. In 2003, Comcast signed a deal to take over pro sports programming in Chicago, starting in the fall of 2004, giving it a lock on league play in three of the nation's top five metro markets.

CHAPTER 14

1. For two views of Malone's career: Mark Robichaux, *Cable Cowboy*, a sober and detailed appraisal that balances Malone's brilliance and flaws but perhaps errs in overstating Malone's importance; and L. J. Davis, *The Billionaire Shell Game: How John Malone and Assorted Corporate Titans Invented a Future Nobody Wanted* (New York: Doubleday, 1998), a cheerfully and relentlessly skeptical depiction of Malone as a selfish cynic, hoodwinking the likes of Barry Diller and Bell Atlantic's Raymond Smith with visions of 500 channels and two-way TV, the author says, which Malone himself had no intention of using for anything but

getting rich. This he accomplished in the sale of TCI to AT&T for $60 billion the year after Davis's book was written.

Malone famously chose to focus on buying chunks of sports, shopping, and science channels, including QVC, through Liberty Media, the piece of TCI he kept. He declared he'd had enough of testifying before committees and getting deposed by lawyers. Or maybe he decided that content matters more than wiring, after all.

2. Robichaux, *Cable Cowboy*, 107.

3. Robichaux, 135.

4. Edward Breen, interview by the author.

5. Robichaux, 215.

6. Ibid., 212.

7. David Lieberman, "Father-son Odd Couple Make Bid to Rule Cable; One's Calm, One's Wary, Together They're Comcast," *USA Today*, 23 July 2001.

8. Robichaux, 209.

9. Frank Drendel, interview by the author.

10. Harold Fitzgerald "Gerry" Lenfest, interview by the author.

11. Drendel.

12. Ibid.

13. Robichaux, *Cable Cowboy*, 209.

14. Bill Gates, interviewed by Charlie Rose at the New York Public Library, March 4, 1998. A transcript of the conversation has been posted on Microsoft's Web site at www.microsoft.com/billgates/speeches/gates-rose.asp.

15. Ibid.

16. Ibid. Microsoft executive Craig Mundie, who was present, later gave a different version of the pitch, with Roberts jokingly suggesting, "Hey Bill, you can solve the problem. Why don't you buy 10 percent of the cable industry?" Rebecca Buckman, "Microsoft Cable-TV Foray Is Costly," *The Wall Street Journal*, 14 June 2002.

17. Nancy Moffitt, "Wharton's Cable Guy," *Wharton Alumni Magazine*, Spring 2000.

18. Gates, interview by Charlie Rose.

CHAPTER 15

1. "Comcast President Brian L. Roberts Testifies about Benefits of Comcast–AT&T Broadband Merger before U.S. Senate Subcommittee," Comcast press release, 22 April 2002.

2. Comcast investors' conference call, 11 February 2004, author's notes.

3. Brian L. Roberts, speech to the Economic Club of Washington (D.C.), 24 January 2001.

4. As presiding director of the Walt Disney Company, Mitchell was charged with responding to Roberts's hostile takeover attempt in February 2004. He quickly spurned the offer, instead rallying to his embattled boss, Disney chairman Michael Eisner. Defending Eisner against dissidents within and Comcast without, Mitchell sounded about as clear as Alan Greenspan on a foggy day: "My belief is that the specific recommendation with respect to Mr. Eisner is based upon a perception that never fully existed as described," Mitchell said, according to an Associated Press report, adding, "But to the extent that it did exist, it has

been completely changed and does not reflect the current reality" (Gary Gentile, "Eisner Keeps Focus on Disney Achievements," 13 February 2004).

The text of Brian L. Roberts's speech to the Economic Club of Washington (D.C.), 24 January 2001, can be read at www.economicclub.org/Pages/archive/fulltext/arch-robertsbl.htm.

5. General Accounting Office, *Issues Related to Competition and Subscriber Rates in the Cable Television Industry*, 24 October 2003.

6. Federal Communications Commission, *In the Matter of Annual Assessment of the Status of Competition in Markets for the Delivery of Video Programming, Eighth Annual Report,* 27 December 2001.

7. Patricia Horn and Ken Dilanian, "Where Connections Are Key to Connecting," *The Philadelphia Inquirer*, 28 January 2001.

8. Edmund Sanders, "Comcast Country Tough on Intruders," *The Los Angeles Times*, 16 July 2001.

9. Paul Schwartzman, "Cable Firm Pulls Plug on Deal in Maryland," *The Washington Post*, 27 August 2001.

10. H. F. Lenfest, interview by the author.

CHAPTER 16

1. Connie Bruck, *Master of the Game: Steve Ross and the Creation of Time Warner* (New York: Simon & Schuster, 1994). Ross "was a person who refused to acknowledge limits; it was this quality that made him at once so magnetic and so flawed," Bruck wrote. "He was daring and bold, and he saw with fresh eyes. But, feeling exempt from legal, ethical, and moral standards, he abrogated them at will."

2. "Cable Godfather," *Cablefax*, PBI Media/Gale Group, 22 May 2003. "An A-list dinner if there ever was one," wrote *Cablefax*'s Paul S. Maxwell on May 19, reviewing the guest list.

3. George Rush and Joanna Malloy, item in column, "Judge Ties Up Assets of 'Matrix' Maker," *The New York Daily News*, 22 May 2003.

4. Comcast annual report; Comcast Foundation federal tax filings, 1999–2002.

5. Comcast–Spectacor Foundation federal tax filings, 1999–2002.

6. *Giving USA 2003,* report prepared by the Center on Philanthropy at the University of Indiana, published by the AAFRC Trust for Philanthropy, the educational and research arm of the American Association of Fundraising Counsel, Glenview, Illinois.

7. Dan Gehringer, "Resurrection," *Philadelphia Daily News*, 30 April 2002.

8. Geoffrey Melada, "Learning Center Brightens North Philadelphia's Future," *Jewish Exponent*, 23 January 2003. "Why do you choose to help non-Jews?" the reporter asked Mrs. Honickman. "Humanity," she said, "is a club bigger than any religion." The Robertses were not quoted in this story or Gehringer's (see preceding note).

9. "The Executive Suite," *The Philadelphia Inquirer*, 14 September 2003, article by the author.

10. H. F. Lenfest, interview by the author. An unusual example of a fortune being used to hold a multigenerational family together is the $100 million Raskob Foundation for Catholic Activities in Wilmington, Delaware. John J.

Raskob was the top financial officer of the DuPont Company in the 1920s and 1930s. He sired a large family, but left the bulk of his fortune to a charitable foundation administered by a professional staff reporting to a committee of Raskob's adult descendants, who hold quarterly and annual meetings at a former family home to give the money away.

11. Peter Binzen, "Recruiter Seeks Executives Who Are Smart—and Nice," *The Philadelphia Inquirer*, 22 June 1998.

12. Tom Infield, "Rendell Backs Donor's Son for Judgeship; The Comcast Chief Has given $17,500 to the Mayor's Campaigns," *The Philadelphia Inquirer*, 18 December 1998.

13. Sural Corporation (Suzanne and Ralph Roberts's investment company), press release, 11 June 1998.

CHAPTER 17

1. Jeffrey Bliss and Jonathan Cox, "Cable, Internet Companies Lobby at Convention over Web Access," Bloomberg LP, 3 August 2000.

2. Leslie Wayne, "Telecommunications Show May Eclipse G.O.P.," *The New York Times*, 2 July 2000.

3. Patricia Horn, "Cashing in on the GOP Convention: Comcast Branded the Gathering in Every Way It Could," *The Philadelphia Inquirer*, 6 August 2000.

4. Murray Dubin, "Republicans Who Puff Have to Do It Outside the Convention Hall," *The Philadelphia Inquirer*, 1 August 2000.

5. Jane M. Von Bergen and Patricia Horn, "It's a G.O.P. Convention but a Comcast Event," *The Philadelphia Inquirer,* 28 May 2000.

6. Thomas Klier and William Testa, "Location Trends of Large Company Headquarters During the 1990s," *Chicago Fed Letter*, Federal Reserve Board of Chicago, June 2002.

7. Later, as governor, Rendell gave no sign he had learned by this experience; as his successor, John Street, struggled with budget cuts and Comcast offered to buy Disney for $66 billion, Rendell as Pennsylvania's governor moved to give developers a tax break to build Comcast a new downtown headquarters—at the city's expense. Rendell justified his give-to-the-rich plan by claiming Comcast would move 1,000 or more jobs into the city, at a time when Comcast was telling investors it would leave Disney in California and had no plans to staff up in Philadelphia.

8. Robert Novak, syndicated column, 31 July 2000.

9. Fox News Network, "Your World with Neil Cavuto," interview with Brian L. Roberts, 28 July 2000.

10. Jonathan Cox, "Comcast Gains Republican Access by Producing Cable TV Coverage," Bloomberg LP, 2 August 2000.

11. Since 2000, Comcast's CN8 news operation has beefed up its permanent staff, which offered gavel-to-gavel coverage of the 2004 presidential conventions.

12. Bliss and Cox, "Cable, Internet Companies Lobby."

13. Seth Hettena, "Congressman Saluted for the Company He Regulates," Associated Press, 3 August 2000.

14. Todd Richissin, "Playing Nice with the Big Money Guys," *The Baltimore Sun*, 2 August 2000.

15. Hettena.

16. Horn, "Cashing in."

17. Ibid.

18. Brody Mullins, "Cable Guy Goes G.O.P.? Companies Look to Shed Democratic Image on K Street," *Roll Call*, 4 March 2003.

One of Comcast's best Republican friends even before its GOP lobbyist–hiring spree has been Sen. Arlen Specter, R-Penn., a hard-edged, former prosecutor, whose son was a schoolboy friend of Brian's. Specter's assignments include the Senate Judiciary Committee's antitrust subcommittee, which periodically fulminates against big corporate mergers like the ones that concentrated the media business in a shrinking number of larger companies in the 1990s. Despite steady financial support from Comcast and other cable companies, Specter sometimes took the floor to pronounce himself concerned about media merger trends, and even hinted darkly about serving Brian Roberts and other cable executives with subpoenas when they failed to show at a 2001 hearing to discuss potential media monopolies. Yet the following April, when the Robertses and AT&T Chairman Michael Armstrong finally came to Capitol Hill to defend their particular merger before the subcommittee, Specter gave his Comcast constituents a warm and friendly introduction to his colleagues, who proved an agreeable audience.

19. The quote about "not losing any sleep" was widely reported; see, for example, Richard Norton-Taylor and others, "Saddam Survived Attack on Building Say British Intelligence Sources," *The Guardian*, 9 April 2003.

20. Brian L. Roberts, Comcast press release, 15 December 2003.

21. Akweli Parker, "Comcast's Growth Produces an Asset: Political Clout; It's Using Its Power to Move the Industry—and Woo D.C.," *The Philadelphia Inquirer*, 6 July 2003.

22. Michael Dressler, "Comcast Dismisses Bereano as Lobbyist to General Assembly," *The Baltimore Sun*, 19 July 2003.

23. Christopher Stern, "Comcast Makes Its Play; Advancing with a Flurry of Acquisitions, the Cable Firm Prepares to Reap the Benefits of a Customer 'Super-Cluster,'" *The Washington Post*, 28 August 2000.

24. Henry J. Holcomb, "When a Philadelphia Tax Break Is a Turnoff; Owners of Center City Skyscraper Shut Off Lights to Protest a Proposal to Aid a New Comcast Corp. Office Tower," *The Philadelphia Inquirer*, 4 June 2004.

CHAPTER 18

1. Don Steinberg, "CEO of New Company Believes in 'Win-win' Deals," *The Philadelphia Inquirer*, 14 November 2002.

2. Don Steinberg, "A Cable Kingdom 40 Years in the Making," *The Philadelphia Inquirer*, 3 November 2002. Malone was "grinning through a clenched jaw" when he delivered that summation to Roberts, at the Denver opening of the Cable Center and its Hall of Fame, Steinberg writes.

3. AT&T press release, 20 October 1997.

4. Ibid.

5. Kris Hudson, "Executive Departs AT&T," *The Denver Post*, 11 July 2001.

6. AT&T press release, 20 October 1997.

7. Mark Robichaux, in *Cable Cowboy*, describes the beginning of Malone's relations with Armstrong; so does Stephen A. Keating in *Cutthroat* (Boulder,

Colo.: Johnson Books, 1999), perhaps the best history of the cable industry, its deals and its personalities, up to the time of its publication.

8. Don Steinberg, "Comcast CEO Is Ambitious but Low-key," *The Philadelphia Inquirer*, 15 February 2004.

9. The notion that Comcast could buy AT&T "was planted in [Brian] Roberts' head" by Microsoft vice president Hank Vigil at a Lindy's delicatessen near Pennsylvania Station in New York, according to Andrew Ross Sorkin and Seth Schiesel, "How the Comcast Deal Came Together," *The New York Times*, 21 December 2001. "Brian, your destiny is to acquire AT&T Broadband," Vigil told Roberts, according to the *Times*. "If that comes up, we'll help you."

10. Akweli Parker and Wendy Tanaka, "Comcast's Work Has Just Ended, Yet Just Begun," *The Philadelphia Inquirer*, 23 December 2001.

11. William C. Smith, "Building a Cable Behemoth," Arthur Block profile, *National Law Journal*, 26 May 2003.

12. Rebecca Blumenstein, "Sweet Revenge: Bid for AT&T Cable," *The Wall Street Journal*, 10 July 2001.

13. David Lieberman, "Father-son Odd Couple Makes Bid to Rule Cable," *USA Today*, 23 July 2001.

14. Christopher Stern, "A Rare Miss for Roberts, Comcast's Dealmaking Executive Surprised by Disney's Rebuff," *The Washington Post*, 29 April 2004.

15. Wendy Tanaka, "A Personal Sales Pitch to Wall Street," *The Philadelphia Inquirer*, 10 July 2001.

16. Brian L. Roberts, Deutsche Bank Tenth Annual Media Conference, 4 June 2002.

17. Ibid.

18. "Comcast's Bid for AT&T Cable," press release, Center for Digital Democracy, 9 July 2001.

19. For excerpts from Senator Boxer's letter, see, for example, David Enrich, "Comcast under Fire for Bundling Practices," States News Service, 28 March 2003. For Senators DeWine and Kohl, see, for example, "Hearing of the Antitrust, Competition, and Business and Consumer Rights Subcommittee of the Senate Judiciary Committee; Subject: Oversight of the Enforcement of the Antitrust Laws," transcript, Federal News Service, Inc., 19 September 2002. The senators' letters were posted on their Web sites and distributed to reporters.

20. Andy Kessler, "Comcast's Pyrrhic Victory," *The Wall Street Journal*, 24 December 2001.

21. Interviews by the author; correspondence between Thomas Creighton and Thomas R. Nathan, Comcast senior vice president for regulatory affairs.

22. Robin Ajello, editor, "Comcast Rides the Broadband," *Business Week*, 11 November 2002.

23. Blumenstein, "Sweet Revenge."

24. Ajello, "Comcast Rides the Broadband."

CHAPTER 19

1. Brian L. Roberts, question and answer session, Citigroup Smith Barney Entertainment, Media and Telecom Conference, Arizona Biltmore Resort & Spa, 7 January 2004, transcript.

2. Ibid.

3. Steve Burke, Morgan Stanley investors' conference, 8 September 2003.

4. Comcast fourth-quarter 2003 report to the Securities and Exchange Commission, 11 February 2004.

5. Roberts, Citigroup Smith Barney.

6. Mark Cooper, "Cable Mergers, Monopoly Power and Price Increases," *Consumers Union*, January 2003.

7. Federal Communications Commission, *In the Matter of Annual Assessment of the Status of Competition in Markets for the Delivery of Video Programming*, Ninth Annual Report, 31 December 2002.

The tenth FCC annual report on competition in video markets, released in early 2004, painted the usual mixed picture. According to the FCC, "The vast majority of Americans enjoy more choice, more programming and more services than any time in history." But at the same time, "cable rates have risen significantly," up 53 percent in ten years, more than double the 25 percent increase in the Consumer Price Index. And while satellite TV had gained more than one-fifth of the pay TV market in the previous ten years, once-promising competitors like phone-based, wireless, and new "overbuild" cable lines had so far "failed to materialize."

8. Roberts, Citigroup Smith Barney.

9. David Lieberman and Andrew Backover, "Qwest, Comcast Duel over DSL Ads," *USA Today*, 28 April 2003.

10. John Curran, "Comcast Refuses Anti-war Ads," Associated Press, 28 January 2003.

11. J. D. Power and Associates, "2003 Residential Cable/Satellite TV Customer Satisfaction Study," report on cable and satellite rates and customer service, 19 August 2003.

12. Thomas W. Watts, "Will Satellite TV's Growth Run Out Soon?" securities research report, S. G. Cowen & Company, 18 June 2003.

13. Cade Metz, "Broadband Scorecard Reports on Nine Broadband Providers," *PC Magazine*, 27 August 2003.

14. Statement by William Cardinal Keeler; this and other quotes were reported by Aparna H. Kumar, Religion News Service, 9 December 2002.

CHAPTER 20

1. Tom Lowry, Amy Barrett, and Ronald Grover, "A New Cable Giant," *Business Week*, 18 November 2002.

2. Q&A with DirecTV's Chase Carey, *BW Online*, 19 January 2004.

3. Ronald Grover and Tom Lowry, "Rupert's World," *Business Week*, 19 January 2004.

4. Mike Farrell and Ted Hearn, "Murdoch Gets DirecTV, But FCC Approval Affords Cable Insurance against Abuses," *Multichannel News*, 5 January 2004.

5. Brian L. Roberts, Comcast Investors' Day, 16 May 2003, author's notes.

6. Comcast investor presentation, 8 September 2003, author's notes.

7. Brian L. Roberts, presentation to Smith Barney Citigroup Entertainment, Media and Telecom Conference, Arizona Biltmore Resort & Spa, 7 January 2004, transcript.

8. "Comcast's Brian L. Roberts Named One of America's Best CEOs by *Institutional Investor*; Roberts Is Top Vote-Winner in Annual Survey," press release, *Institutional Investor*, 16 January 2004. According to the release, "Roberts received more votes than any other CEO in a survey to which 1,374 portfolio managers and securities analysts managing about $4.5 trillion at 405 investment firms responded."

The article noted Comcast's emphasis on team leadership, the depth and breadth of executive experience within the company, and the success of the recent acquisition and integration of AT&T Broadband. The magazine praised Comcasts's "stellar results" and efficient growth and, with the acquisition of AT&T Broadband, the company's transformation into "the undisputed industry leader."

9. Jea Shim, "A Q&A on the Ebb and Flow of Broadband Competitive Dynamics; While Price War Is Unlikely, Cable HSD Growth Should Slow in Coming Quarters," securities research report, Tradition Asiel Securities, 10 February 2004.

10. Richard Morgan, "Stargazing," *The Daily Deal*, 5 January 2004.

11. The *Washington Post* went even further in its provincialism, identifying Brian Roberts, head of Washington's biggest cable provider, as "the son of a Philadelphia beltmaker" (Frank Ahrens, "An Audacious Attempt at Media Giant Status," 12 February 2004). Ralph had owned a belt company in the 1950s, but he was more often identified as Comcast's founder and chairman, the posts he had held for forty years.

12. Brian L. Roberts, Comcast investors' conference call, 11 February 2004, transcript and author's notes.

13. Akweli Parker, "Comcast to Pitch to Big Disney Investors," *The Philadelphia Inquirer*, 13 February 2004.

14. Comcast conference call, 11 February 2004, transcript and author's notes.

15. Saul P. Steinberg ran Reliance for thirty years, paying himself, his family, and his (and the Robertses') alma mater, the University of Pennsylvania's Wharton School, many millions before he ran out of cash, leaving Reliance bondholders, banks, major policyholders, and citizen-funded insurance bailout funds an estimated $3 billion in the red. That was in 2001. Two years later, the university passed a resolution praising Steinberg for "demonstrating the power of the individual within a free-market system" and for "sharing his extraordinary vision, energy, and enthusiasm, and the fruits thereof, with the University." Shareholders, bondholders, and the Pennsylvania Department of Insurance have sued Steinberg. Wharton instructors occasionally use Steinberg deals as case studies.

16. Brian L. Roberts, Comcast conference call, 11 February 2004, transcript and author's notes.

17. *Larry King Live*, Cable News Network, February 20, 2004.

18. George Mitchell, introduction to Brian Roberts's speech to the Economic Club of Washington (D.C.), January 24, 2001; see also Chapter 15, footnote 4.

19. Fulcrum Global Partners, update report on Comcast–Disney proposal, www.fulcrumgp.com, 26 February 2004.

20. Wendy Tanaka, "Comcast Bid May Shield Eisner," *The Philadelphia Inquirer*, 5 March 2004.

21. Brian L. Roberts, Bank of America media conference, New York, 20 May 2003.

22. Ronald Grover, "How Comcast Let the Mouse Get Away," "Power Lunch" column, *Business Week*, 29 April 2004, online edition.

23. "Comcast to Offer Net Phone Service: Cable Company Plans Aggressive Roll Out of VoIP Service," Reuters, 26 May 2004.

24. See, for example, "Talk Gets Cheap," *Barron's*, 24 May 2004; Andrew Kantor, "Using the Net as a Telephone Service Finally Sounds Reasonable," column, *USA Today*, 5 March 2004; Alex Salkever, "These Phone Calls Aren't Phone Calls," *Business Week*, 13 February 2004.

25. David Lieberman, "Comcast Poised to Offer Phone Service—Just Not So Fast," *USA Today*, 27 May 2004.

26. Chitra Somayaji, "Comcast's Roberts Named Chairman," Bloomberg LP, 26 May 2004.

27. Thomas W. Eagan, "VoIP: An Improved Armstrong Strategy," research report, Oppenheimer and Company, New York, 26 May 2004.

28. Mark Jaffe, "Comcast Fails Governance Standards Criticized at Walt Disney," Bloomberg LP, 16 February 2004.

29. CWA published its count on www.ComcastVoteNo.com.

30. I am indebted for the brief account of the union protesters to Akweli Parker of *The Philadelphia Inquirer*.

31. Declan McCullagh, "Comcast: We're the Biggest Spammer on the Internet," www.CNETnews.com, 24 May 2004.

32. Sallie Hofmeister, "E! Chief's Reign May Be Ending," *The Los Angeles Times*, 27 May 2004.

33. "Comcast to Crack Down on Cable Thieves," *South Florida Business Journal*, 11 May 2004.

34. www.BroadbandReports.com, 12 May 2004.

35. David C. McCourt, "Why I'm Filing Chapter 11," *The Wall Street Journal*, 21 May 2004.

36. Howard J. Blumenthal, interview with the author.

37. Brian L. Roberts, see Footnote 21, above.

SELECTED BIBLIOGRAPHY

Although Comcast has not previously been the main subject of a commercially published book, its development over the past forty years has been reported in many articles in national, regional, business, and trade periodicals, and in securities research reports by Wall Street investment houses. In addition, Ralph and Brian Roberts and their lieutenants have been portrayed anecdotally in a number of works about better-known (though perhaps less important) media figures.

In particular, archival stories from *The Philadelphia Inquirer* figure prominently in this book.

Below is a list of other major sources cited.

PERIODICALS, NEWSPAPERS, AND NEWS SERVICES

Associated Press
The Baltimore Sun
Barron's
Bloomberg LP
Business Week
Cable News Network
Cable World
Chicago Fed Letter
Crain's Electronic Media
The Daily Deal
The Denver Post
Federal News Service
Fortune
The Guardian
The Los Angeles Times

Multichannel News
New Rochelle Standard-Star
The New York Daily News
The New York Review of Books
The New York Times
PC Magazine
Philadelphia Daily News
Religion News Service
Reuters
Roll Call
Sports Illustrated
States News Service
USA Today
The Wall Street Journal
The Washington Post

BOOKS, REPORTS, AND A SAMPLING OF OLDER ARTICLES

Aaron, Daniel, with David A. Long. *Take the Measure of the Man: An American Success Story*. Philadelphia: Veritas Publishing, 2001.

Brown, Kathi Ann. *Wired to Win: Entrepreneurs of the American Cable Industry*. Fairfax, Va.: I. Martin Pompadur, Spectrum Publishing Group, 2003.

"Cable Television: New Punters in the Cultural Wasteland." *The Economist*, 25 October 1980.

Center on Philanthropy at the University of Indiana. *Giving USA 2000*. Glenview, Ill.: AAFRC Trust for Philanthropy, 2003.

Cooper, Mark. "Cable Mergers, Monopoly Power and Price Increases." *Consumers Union*, January 2003.

Eagan, Thomas. "VoIP: An Improved Armstrong Strategy." Securities research report. Oppenheimer and Company, New York, 26 May 2004.

Federal Communications Commission. *In the Matter of Annual Assessment of the Status of Competition in Markets for the Delivery of Video Programming.* Eighth, ninth, and tenth annual reports. Washington, D.C.: GPO, 2001, 2002, 2004 (respectively).

General Accounting Office. *Issues Related to Competition and Subscriber Rates in the Cable Television Industry.* Washington, D.C., 24 October 2003.

Hundt, Reed E. *You Say You Want a Revolution: A Story of Information Age Politics*. New Haven: Yale University Press, 2000.

Keating, Stephen A. *Cutthroat*. Boulder, Colo.: Johnson Books, 1999.

Power, J. D., and Associates. "2003 Residential Cable/Satellite TV Customer Satisfaction Study." 19 August 2003.

Robichaux, Mark. *Cable Cowboy: John Malone and the Rise of the Modern Cable Business*. Hoboken, N. J.: John C. Wiley & Sons, 2002.

Senate Judiciary Committee. Antitrust, Competition, and Business and Consumer Rights Subcommittee. *Oversight of the Enforcement of the Antitrust Laws: Hearing of the Antitrust, Competition, and Business and Consumer Rights Subcommittee of the Senate Judiciary Committee*. Transcript. Federal News Service, 19 September 2002.

Shim, Jea. "A Q&A on the Ebb and Flow of Broadband Competitive Dynamics; While Price War Is Unlikely, Cable HSD Growth Should Slow in Coming Quarters." Securities research report. Tradition Asiel Securities, 10 February 2004.

Sloan, Allen. "Bring Plenty of Money." *Forbes*, 10 December 1979.

"Video's Effect on Political Campaigning." *TV Digest*, 14 June 1952.

Watts, Thomas W. "Will Satellite TV's Growth Run Out Soon?" Securities research report. S. G. Cowen & Company, 18 June 2003.

ORGANIZATIONS AND WEB SITES

The Benton Foundation, **www.benton.org**
Broadband Reports, **www.broadbandreports.com**
The Cable Center, **www.cablecenter.org**
Center for Digital Democracy, **www.democraticmedia.org**
CNET News, **www.CNETnews.com**
Comcast Corporation, **www.cmcsa.com**
Communications Workers of America, **www.cwa-union.org**
Federal Communications Commission, **www.fcc.gov**
Fulcrum Global Partners, **www.fulcrumgp.com**
Internal Revenue Service, **www.irs.gov**
Microsoft Corporation, **www.microsoft.com**
Securities and Exchange Commission, SEC Edgar database, **www.sec.gov**